FORDHALL FARM
The Yoghourt Years

Copyright © Connie Hollins, Fordhall Farm 2015

ISBN 978-0-9930073-2-3

Ellingham Press, 43 High Street, Much Wenlock,
Shropshire TF13 6AD
www.ellinghampress.co.uk

Designed by MA Creative Limited www.macreative.co.uk

Front cover image: Young Marianne Hollins eating a Fordhall yoghourt

FORDHALL FARM
The Yoghourt Years

Edited by Charlotte Hollins

May Hollins

Ellingham Press

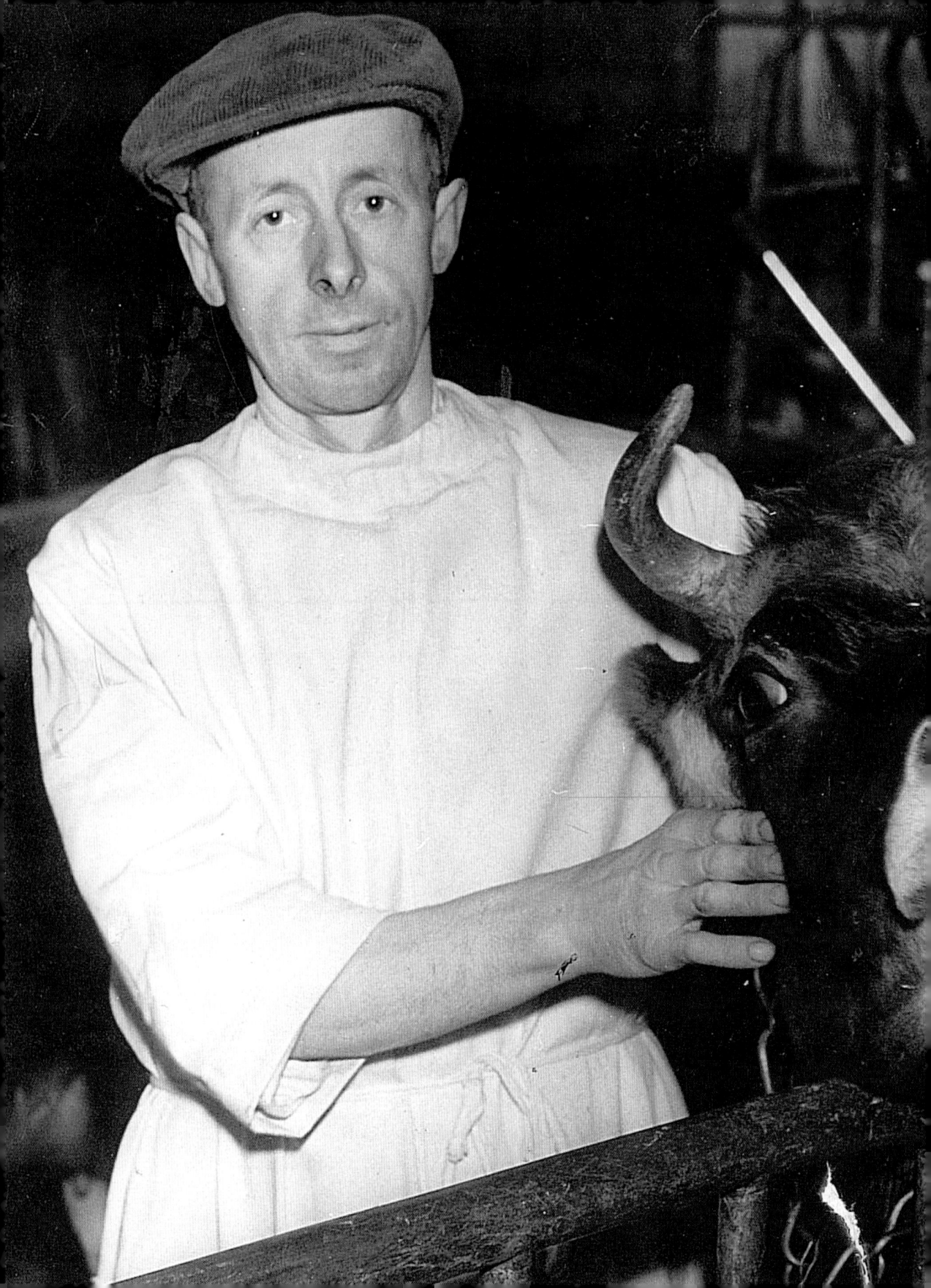

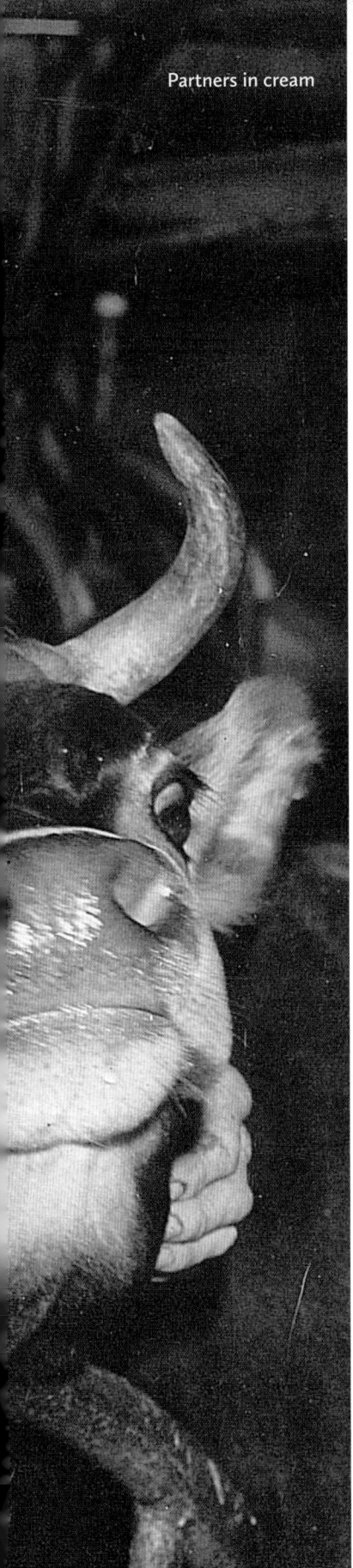

Partners in cream

Introduction

Fordhall Organic Farm is situated in a beautiful part of north Shropshire, adjacent to Market Drayton and along the Tern Valley. Bounded by the River Tern and the main road from Shrewsbury to Newcastle, this island farm has a strong base in the area's local history.

Today it is famous for its unique community ownership structure and long organic heritage, but few people know that Fordhall is also recognised as one of England's very first commercial yoghourt producers.

Yoghourt production began at Fordhall in the late 1950s, led by the pioneering May and Arthur Hollins. Their love and appreciation for Mother Nature, natural foods and good health, took them on a journey of discovery and innovation. The result was an explosive invention of dairy products which were sold in their thousands across the length and breadth of Britain.

Although Fordhall no longer produces yoghourt, or indeed any dairy products, the organic ethos thrives at the farm to this day. This is a story of triumph, hard work, dedication, innovation and a fight for survival.

This book was generously funded by the Heritage Lottery Fund. It forms just one element of the Project's final objectives: there is a much fuller Archive available in Shrewsbury, as well as more stories, memories and articles on the Fordhall Farm website. An interpretation panel can also be found on the farm, allowing visitors to listen to the real memories of those who experienced the yoghourt years at Fordhall.

A committed team of staff and volunteers at Fordhall Farm has researched the facts and stories contained in the following pages. They have delved through the many papers held at Shropshire Archives and the newspaper cuttings held at Fordhall Farm, not to mention hundreds of photographs, old diaries and promotional material. Most importantly, they have listened to and gathered the memories of those who worked at the farm at the time, allowing this story to be brought into the public domain.

We hope that the book will act as a reference for local historians, an interesting read for the coffee table, a memory for those involved, and a reminder that one of the mainstays of our weekly shop was only introduced a relatively short while ago.

Charlotte Hollins

Fordhall Farm *The Yoghourt Years*

Early Days

Arthur and his son Robert with Jersey calves

Arthur's Mother, Lillian Hollins

Arthur Hollins was born in 1915, when his family had been working the land at Fordhall Farm for several generations. The farm, on the outskirts of Market Drayton, started out as 150 acres of grazing land with sandy slopes and water meadows leading down to the River Tern. Arthur's father Alfred had managed Fordhall as a mixed farm, growing vegetables and keeping some livestock, including a small herd of dairy shorthorn cattle.

After some years of ill health, Alfred died in 1928 aged only 38, leaving the farm tenancy to 13-year-old Arthur and his mother Lillian. Unfortunately, in the push to grow more food during and after the 1914-18 war, Alfred had relied heavily on artificial fertilisers which gradually depleted the natural fertility of the soil, leaving a legacy of decreasing yields and spiralling debt. Between them, Arthur and Lillian struggled to make a living for a number of years, selling whatever surplus they managed to produce at the local market. This included Lillian's delicious Cheshire cheese and cream cheese, which always proved popular.

In 1942, Arthur married May Baker, a Land Army girl from Birmingham whom he met at a farmers' dance. Three children followed within the next ten years, all of whom were involved in the later success of Fordhall Farm.

Arthur's father, Alfred, on the right enjoying a break with the farm workers

Arthur and May

Fordhall Farm *The Yoghourt Years*   3

Like all wartime farmers, Arthur had only been allowed to market a limited range of products. It was illegal to supply cream for sale and cheese could only be made using a national recipe, nicknamed "Government Cheddar", to suit the rationing system. This was mainly done in government factories where most of the milk production was sent.

At first, Arthur grew mushrooms commercially, finding they not only provided a much-needed source of income, but also large quantities of spent compost, which helped to improve the condition of Fordhall's soil. Then, Arthur and May decided to set up a country club at the farm for the officers and men at nearby Tern Hill RAF station. They had already been obliged to build a large water cistern in case of fire at the farm; ingeniously, the opportunistically-minded Arthur had designed and built it as a swimming pool. This became one of the country club's main attractions, along with archery, tennis and other recreational pursuits.

Arthur and May both believed in enhancing the soil's natural fertility without the use of artificial additives; by 1949 Arthur was experimenting with organically grown vegetables, known then as 'compost' vegetables, although there was no ready market for them at that time. They were also keen to establish a pedigree Jersey herd in order to make use of the superior quality milk for which the breed is renowned. However, even by selling the commercial shorthorn herd, they were still short of funds. They knew they would have to work hard to be able to afford such valuable stock, so they set to the task without delay. The country club was re-launched and Arthur and May worked all hours, entertaining guests and encouraging the public onto their little piece of Shropshire.

Fordhall Farm *The Yoghourt Years*

CLUB HOUSE
A large double-windowed old oak-beamed room, pleasantly furnished for old-world comfort, with log fires for Bridge and other Parties, includes a Bechstein Piano and Gramophone. There are other rooms for Table Tennis, Darts, Table Bowls, etc.

LAWN TENNIS
Excellent Grass Court in pretty surroundings.

ARCHERY
Two Full Ranges of 25 and 50 yards set within the Club Gardens; also small Indoor Target.

BADMINTON, TENNI-KOITS, CROQUET, DART-BOWLS
Are set out on the well-kept Lawns.

CHILDREN'S CORNER
Paddling Pool, Sand Pit, See-Saws and Swings under safe surroundings will provide hours of care-free and healthy fun, whilst parents relax, or join in the amenities.

RIDING SCHOOL
Hunters are attached to the Club for Members' use. Lessons given. No country or conditions could be better for the purpose.

CATERING
The Club specialises in "Home-made" Cakes. Teas are served in Continental style in the old-world Garden, surrounded by Lawns, Shrubs and Fountains. Picnics, Lunches, Milk Cocktails, Tobacco, Cigarettes and Minerals are on Sale.

HOLIDAYS AND WEEK-ENDS
Facilities are made available for Members to stay for Holidays and week-ends at the Club to make the fullest use of the Shropshire beauty spots.

WINTER FACILITIES
Skating Lake, Tobogganing Banks, Club Parties, Dances, Bridge (Special Bridge Teas), Hockey.

BIRTHDAY PARTIES, Etc.
For Members and their friends, by special arrangement, with full use of the Club and Grounds.

WALKS
Are interesting and extensive in the Farm Lands and Lanes, and for GOLFERS a very attractive Nine-Hole Course is situated nearby.

A new look for the country club in the 1950s, revived to raise funds for Arthur's Jersey herd

SWIMMING POOL (42 ft. × 16 ft.)
Depth 3 ft. 6 in. rising to 6 ft. 6 in., with three sets of Steps and Spring Diving Boards. The Pool is of Fresh Chlorinated Drinking Water has passed every test and had high praise bestowed for its healthy and pure condition.

Fordhall Farm *The Yoghourt Years*

Jersey cows gave great tasting milk

At first all milking was done by hand

"He did everything there, there was archery at the side of the house, he had his two riding horses, the swimming pool. Every spring, Arthur would turn a valve and it would drain out into the fields; taking about a week. We would go in, scrub it out, take down wheelbarrows and whitewash it all with brooms, and then the hose was turned on and it was filled up again and it was used all through the summer... Beyond that was a sandpit for the children and a crazy golf course, and in the field on the side they did clay pigeon shooting... the tennis court doubled up as a croquet lawn. Inside there was snooker and darts. There were damson trees down the one side where the cattle used to go and they used to make the most beautiful damson jam... I used to pick the watercress from the meadow and May used to make beautiful watercress soup, there was always something going on." **Mary Cowen**

The Jersey Herd

May (right) and dairymaid with Jersey calf

The Jersey Herd

By 1951, after two intense years of effort on both the farming and hospitality fronts, Arthur and May were finally able to buy their first pedigree Jersey stock. Following a fact-finding family holiday on Jersey, a small herd of hand-reared cows was shipped to Shropshire; the first step on the Fordhall dairy journey.

As Arthur explained in his book, **The Farmer, the Plough and the Devil,** Jersey cows were a natural choice:

> "Their milk was much richer and contained more butter fat than our shorthorns provided. Also, despite their size, their yield of milk was much higher. Since we were processing much of the milk ourselves, these would, we believed, be tremendously advantageous."

Starting with the original 14 cows, the herd gradually increased in size over the years, building up to 100 cattle by 1961. Compulsory testing for bovine tuberculosis was introduced in the early 1950s, and Fordhall's cattle were declared TB-free which meant the herd could be advertised as TT (Tuberculin Tested). In order to bring the

"There was one cow that was only milking two pints a day and you'd say 'Gaffer, is there any point in keeping her?' and he'd say 'foundation stock' – he was very fond of the animals, knew them all, he rarely milked, but he knew all the animals and foundation stock weren't to be culled. They were originals and they had to stay even though they weren't producing anything."

Geoff Fletcher

Arthur standing proudly with Jersey cow

8   Fordhall Farm *The Yoghourt Years*

Cattle graze through the snow, pulling up the dead grass, which acts as roughage to keep them warm

The Jersey herd heads for the milking parlour

dairy operation up to date, Arthur invested in a new milking parlour and a milking machine. He was also experimenting with keeping his stock outside all winter, a revolutionary idea at the time. At first he tried keeping only half the herd outside, but as he later explained in a report in the **Staffordshire Sentinel** in 1957, *"I found they gave the same average yield and in fact they were healthier and went to the spring grass better than those kept inside."* As a result Arthur's herd remained outdoors all year round.

Despite their hard work and innovation in developing the Jersey herd and farming system, Arthur and May were still only allowed to sell liquid milk until the restriction on the sale of other dairy products was finally lifted with the end of rationing in 1954. But with their preparations in place, they were able to make an immediate impact on the milk and cheese market.

Fordhall Farm *The Yoghourt Years*   **9**

A myriad of milk products

May with Fordhall Jersey clotted cream

CLOTTED CREAM

Free to develop their Jersey milk products following the end of rationing, Arthur and May set to work, and in 1956 discovered how to make clotted cream in the traditional manner. May's niece, Mary Cowen remembers:

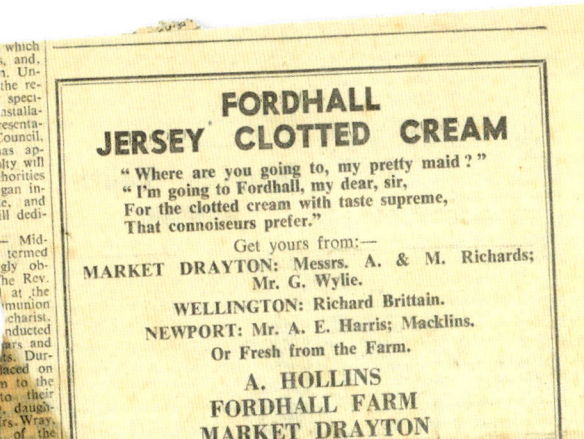

"While I was there (at home in Cornwall) Arthur came down on holiday and he asked me if I could take him to some farms that made clotted cream. We looked up quite a few and visited them. They were quite happy to tell him their secrets. Clotted cream was only made in Devon and Cornwall then, but Arthur went back to Fordhall and started to make it there."

Fordhall clotted cream was an instant success, and within two years the farm was producing 4,000 cartons a week. Selling was mostly local at first, but demand was there even then, and it was a logical step to use the clotted cream unit to produce more dairy specialities.

Before long there were Fordhall Farm stalls at a number of neighbouring markets and in towns as far away as Liverpool, selling a superb array of clotted cream, farmhouse butter, Jersey TT milk, Jersey cream, buttermilk and farmhouse Cheshire cheese. So soon after the end of rationing, customers must have felt rather like a child in a sweetshop enjoying delicious mouth-watering, fresh food made on the farm.

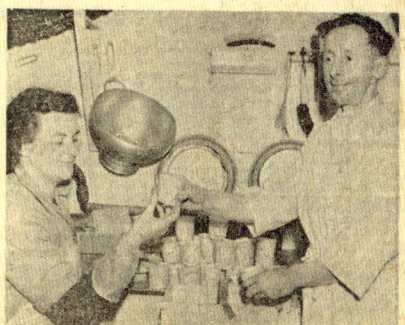

Newspaper cuttings from 1956

Fordhall Farm *The Yoghourt Years*   11

CREAM CHEESE
(also known as Soft Cheese)

Before Arthur invested in his Jersey herd, his mother had for a long time made cream cheese for sale on the local market. Arthur started to produce the cheese again using the skimmed milk left over from making clotted cream. Mary Cowen describes the process below:

> "They used to separate the cream with a cream separator, and Arthur started to make curds with this skimmed milk. He built a whole pile of little wooden boxes about nine inches square divided down the middle, lined them with muslin and poured the curds in to drain. When it was drained you pulled out the muslin, wrapped up the cheese, wrapped it in greaseproof paper and that was ready to be sold as soft cheese."

At first, the cream cheeses were simply flavoured with parsley, onion and walnuts. By 1965, as this quote from **Delicatessen** magazine shows, there were some quite exotic varieties:

> "There are more than 20 varieties of Fordhall flavoured cheeses. The flavourings include onions, chives, pineapple, horseradish, cucumber, nuts and wine, garlic, pickled walnuts, celery or celery herb, pimento, parsley and herbs, apricot chutney, shrimp, lobster, smoked salmon and caviar."

Eventually, there were over 40 different flavours, thanks to suggestions from customers, and Arthur and May's flair for experimentation.

1957 May Hollins making cream cheese

12   Fordhall Farm *The Yoghourt Years*

Straining the cheese through muslin

Above: 1961 Cutting the curds away from the whey after the rennet has been added. The curds were then made into cheese'
Below: 1961 Adding watercress to cottage cheese

DELICIOUS, NUTRITIOUS

FORDHALL FARMHOUSE FOODS

FOR BETTER HEALTH

FRESH FROM THE FARM

Fordhall, a cheese making farm for generations, delightfully situated in Shropshire, maintains their herd of Jerseys naturally, without the use of harmful sprays or artificials.

Our Cream, Yoghourt and natural cheeses are produced daily on the farm in our very up to date dairy.

All the ingredients used for flavouring are compost grown and no artificial flavourings are used.

We think this natural approach to food protects your health.

MEMBER OF THE SOIL ASSOCIATION

A. & M. Hollins, Fordhall Farm, Market Drayton, Shropshire
Telephone: TERNHILL 255

Fordhall Farm *The Yoghourt Years* 13

Arthur inspects cream cheese

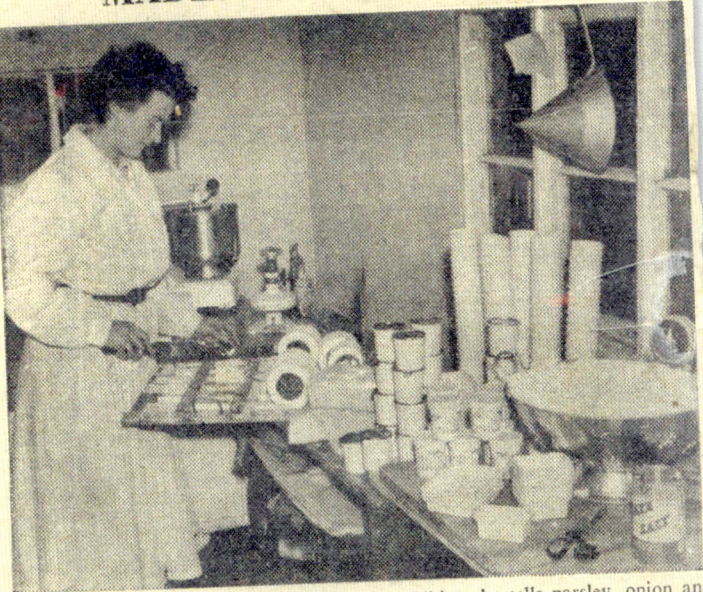

WELLINGTON JOURNAL & SHREWSBURY NEWS, NOV. 16, 1957

FARMHOUSE YOGHOURT BEING MADE IN SHROPSHIRE

DO you want to live to be one hundred? In Bulgaria and many central European countries the shepherds and other country dwellers, many of whom live to a great age, quite a number reaching the century, attribute their health and longevity to Yoghourt made with goats' or cows' milk. Now Yoghourt, developed for hundreds of years in Europe, is being made in Shropshire and this health-giving food, extremely popular on the Continent and in the United States, is fast becoming a favourite in Britain.

Producer of Yoghourt in Shropshire is Mr. A. Hollins, Fordhall Farm, Tern Hill. In his dairy the Yoghart is prepared by adding the pure Yoghourt culture to the still-warm milk. Mr. Hollins told the "Journal" that the health-giving Yoghourt bacteria, injected into the milk, kills all harmful bacteria. Mr. Hollins also produces Yoghourt cheese and his products are now being sold all over central England.

On his 150-acre farm, Mr. Hollins has a pedigree and non-pedigree herd of Jersey cattle numbering 52 head. He started production of Yoghourt and Yoghourt cheese in May of this year and now over 200 gallons of milk per day are being used to make Yoghourt products.

The venture started with clotted cream production, of which Mr. Hollins is one of the few producers north of the southern counties. He sells about 2,000 cartons per week, reaching nearly 4,000 at the height of the strawberry season. In addition, he sells parsley, onion and walnut cheese and Jersey farm butter.

Health value

"There is no question about the health value of Yoghourt," says Mr. Hollins. "The Yoghourt bacteria tones up the system and this is proved by the number of customers who buy it regularly and are loud in their praises of its qualities." This from a man who told the "Journal" that he is the only one in central England producing farm-house Yoghourt from cows' milk and who made it for his own table long before it became a commercial enterprise. Not only can the Yoghourt be taken as a drink, but it can be drunk with fruit juice, eaten with cereals and fresh or dried fruit and in place of custard when sugar is added, or made into salad dressing.

Mr. Hollins' next enterprise is to be the production of cottage cheese. From an American firm he has obtained the recipe of a famous American brand of this product and when he commences production he will be England's only producer of this brand so far as he is aware. The method of making it, he says, is an entirely new one.

To milk his herd — which he did single-handed for four years — Mr. Hollins has shown an inventive turn of mind by adapting a milking machine to enable it to milk four cows at once and carry the milk direct to the separating apparatus. The use of buckets is thus eliminated.

YOGHOURT

May first introduced Arthur to the health benefits of yoghourt when he was suffering from a delicate stomach, and in May 1957 they started to make it commercially, at a time when the herd of Jersey TT cattle numbered 52 head. By November of the same year they were using over 200 gallons of milk a day in yoghourt products as reported in the *Wellington Journal & Shrewsbury News*. The yoghourt was prepared by adding a pure live yoghourt culture (purchased from Europe) to the pasteurised warm milk. Arthur imported the live culture each week until he developed his own home-grown cultures in Fordhall's onsite laboratory.

At first, they sold plain live yoghourt, but soon Arthur began to experiment with various fruit flavours, the first of which was 'pure pineapple juice'.

The list of varieties is extensive, but included more unusual ingredients such as cherry and hazelnuts, coffee, damsons, red wine and tangelo (a cross between a Duncan grapefruit and a Dancy tangerine). Wherever possible, locally grown fruits were used in Fordhall yoghourts, but citrus and tropical fruits were bought from a local greengrocer.

Arthur also made a most unusual yoghourt cheese which was manufactured in the same way as cream cheese. The yoghourt was strained through muslin and the yoghourt whey used elsewhere. What was left was yoghourt cheese, much lighter than standard cream cheese and with a slightly 'tart' taste.

1961 Cutting oranges for flavouring yoghourt

16 Fordhall Farm *The Yoghourt Years*

Above: May and Arthur together on a late 1950s stall
Below: Products were initially packed by hand with foil lids

Fordhall Farm *The Yoghourt Years*   17

"FORDHALL" YOGHOURT

"FORDHALL" YOGHOURT is Milk, produced under the most hygienic conditions and used fresh from the cow, with a pure oriental lactic culture (Bacillus Bulgaris) added at the correct temperature, making it digestible and therefore excellent for young children, invalids and those with weak digestions.

On the continent and in America it is delivered daily with the milk. In this country it is popular with athletes, health-conscious people and, when recommended by the doctor, for Diabetes, Nerves, Acidity, Ulcers, Constipation, Slimming, and after treatment with antibiotic drugs.

"FORDHALL" YOGHOURT tones up the system, corrects the wrong bacteria in the stomach, helps the glands to work and makes you resistant to disease. It is a valuable source of Calcium and Vitamin B.

"FORDHALL" YOGHOURT is the finest health-giving and slimming food in the world. In France Yoghourt is the staple diet. French ladies are noted for their wonderful figures. Our culture comes each week from Paris.

You can now enjoy a continuous supply of this perfect food by following these simple directions, but it is important to obtain a fresh culture ("Fordhall" Live Yoghourt) each week.

Heat one pint of fresh sweet milk to 150°, pour a little into a clean jug and cool to 112°, then add one wineglassful of "Fordhall" Yoghourt and mix well. Cool remainder of milk to 130° and add to mixture in jug, mix well. Place in warm place for at least six hours or overnight. An airing cupboard is excellent.

WAYS OF USING "FORDHALL" YOGHOURT

As a drink with any fruit juice and plenty of sugar—excellent with Ribena or the juice from tinned or bottled fruit.

In place of custard with fresh or dried fruit and plenty of sugar. Or with cere[al]

For cocktail parties it is delightful with sherry and sugar added—something diffe[rent] touch.

Mixed with jam for elevenses or T.V. interval.

Plain, with the addition of brown sugar and thin cream.

SALAD DRESSING.—One carton of "Fordhall" Lemon Yoghourt straight fr[om] the addition of a little Salad Cream.

RICH SALAD CREAM.—One "Fordhall" Sour Cream or Smetana. One [Yog]hourt. Brown sugar to taste. Mix all together. The addition of a little gr[ated] health-giving change.

ICE CREAM.—One carton or more of "Fordhall" Fruit Flavoured Yoghou[rt] carton of "Fordhall" Whipped Cream and freeze.

LACTIC CHEESE CAKES.—½-lb. "Fordhall" Yoghourt Cheese or "Fordha[ll"] tablespoon wholemeal flour, lemon rind, brown sugar, currants, knob of b[utter] all together, put into pastry cases and bake in moderate oven till set (abou[t]

VARIATIONS:—
1. Grate a little nutmeg on top before baking.
2. Slice a banana thinly and add to mixture before baking.
3. Put into pie dish instead of pastry and use as sweet with fresh or ste[wed]
4. Add chopped onions and a little grated cheese to the mixture before ba[king] either with pastry or without.
5. Add ¼-lb. ground almonds, one whisked egg white, and either a lit[tle] before baking.

LACTIC SAUCE.—Cook any fish gently with a knob of butter, chopped [onions] desired. Pour off milk and onions, add one carton of "Fordhall" Lemo[n] over fish, serve with mashed potatoes and tomatoes.

PANCAKES.—Cream together 2-oz. butter and 2-oz. sugar. Add 2 egg[s] warm one carton of "Fordhall" Yoghourt with ¼-pint milk very g[ently] Half fill bun tins and bake in moderate oven for 10–15 minutes. Wh[en] and eat hot with jam.

FORDHALL YOGHOURT CHEESE

This delicious Cheese is made from yoghourt produced with a pure oriental lactic culture, Bacillus Bulgaris, and T.T. Jersey milk (fat-reduced) under the most hygienic conditions at Fordhall Farm, Market Drayton.

For your cocktail parties it makes the perfect base for savoury delicacies, spread on dainty biscuits. It is perfect with any wine.

It may be used as a sweet or savoury with any addition, or as a spread for picnics and packed lunches. Excellent with spring salads.

This old-time slimming and health-producing cheese has been used for thousands of years in the Balkans, and is now becoming more widely recognised in this country. You will benefit most by using it in place of butter.

As a predigested food it is invaluable for growing children; they will love it used as a spread, alone or mixed with sugar, honey, treacle, jam, marmite, etc.

Recommended by the medical profession for gastric troubles, diabetes, delicate children and various diets. It also aids natural elimination of the bowels, and builds up resistance to disease, providing a readily absorbed source of calcium and Vitamin B.

It helps to restore the natural bacterial flora of the system after treatment with anti-biotic drugs.

FORDHALL YOGHOURT CHEESE is a perishable product guaranteed not to contain any preservative, colouring salt or rennet. Best kept in a cool place.

W. & S. Ltd, S'bury. 8765

Due to growing customer interest in their dairy products and after many requests for how to use them in cooking, May decided to provide various helpful leaflets on Fordhall stalls.

Top: 1972 Making cottage cheese
Bottom: Packing cottage cheese

COTTAGE CHEESE

In 1957 Fordhall Farm was producing a whole array of dairy products. However, Arthur was not satisfied and wanted to extend the range even more by making cottage cheese.

As was reported in the **Wellington Journal & Shrewsbury News** of November 16 1957:

> "Mr Hollins' next enterprise is to be the production of cottage cheese. From an American firm he has obtained the recipe of a famous American brand of this product and when he commences production he will be England's only producer of this brand so far as he is aware.
> The method of making it, he says, is an entirely new one."

Fordhall Farm *The Yoghourt Years*   19

YOGICE AND YOGTAILS

By 1967 Arthur and May were producing at least 12 different fresh-fruit yoghourts, but they still had time to develop Yogice. Yogice was a blend of Fordhall live fat-free yoghourt and fresh Jersey cream. It was said to provide a refreshing alternative to modern ice cream along with the extra benefits contained in live yoghourt. The five original varieties on offer were:

- Yogice with fresh vanilla and lavender honey
- Yogice with pure chocolate and fine-ground Swiss hazelnuts
- Yogice with pure emulsified lime
- Yogice with fresh strawberries
- Yogice with wild blackberry liqueur and fresh blackberries

Yogtails were introduced around the same time as Yogice. They were described as *"skilfully blended live fat free yoghourt, citrus fruit, and fresh berried fruits, advocaat, red wine or rum."* Yogtails were advertised as *"a new approach to your party, slimming diet and a mid-morning break, it also makes an excellent pick-me-up"*. The advocaat used in the Yogtails was made on the farm, from free-range eggs, Barbados sugar and brandy.

"On a Sunday, I used to do Yogice. I was only 13 when I started; you earned your pocket money. We loved Yogice and used to eat more than we packed. You pulled this little handle and put the pot underneath, 'cos you did it all by hand. It used to come out like ice cream, and you would twirl it into the pot, like you do a cone for ice cream." **Helen Burgess**

Fordhall Farm *The Yoghourt Years*

NEW FROM FORDHALL

YOGICE

Fordhall are already producing 12 different fresh fruit yoghourts, which are unique in their live tanginess. Now they offer an exciting addition to their dairy products. YOGICE, a delicious blend of Fordhall Live Fat Free Yoghourt and fresh Jersey cream. This provides a refreshing alternative to modern ice cream with the extra health benefits contained in live yoghurt. Every recipe where ice cream is used Yogice is better.

Fordhall offer five varieties:

Yogice with fresh Vanilla and lavender honey.

Yogice with pure Chocolate and finely ground Swiss Hazelnuts.

Yogice with fresh Strawberries.

Yogice with pure emulsified Lime.

Yogice with wild blackberry liqueur and fresh Blackberries.

Pure lavender honey is used with each variety.

Each one is presented in 4 and 5 oz. insulated caftons, to ensure that the texture and coldness is retained for three hours after purchase. When thawed, it makes a delicious whipped dessert with liquid fruit base.

It is wonderful for children or alternatively the whey and fruit may be whipped together to make a delicious creamy yoghourt which can then be refrozen if desired for a cold sweet. Unlike ice cream it is quite correct to enjoy Yogice this way, as it is a live fresh product.

Table wine or fruit juices whipped into the Yogice and folded into a jelly, makes an exciting party treat. Allowed to thaw and laced with whisky, a chocolate Yogice is a wonderful and safe drink for parties.

All varieties make an excellent addition to home-made punches.

A. & M. Hollins, Fordhall Organic Farm, Market Drayton, Shropshire.

All Enquiries Telephone: TERNHILL 255

Member of the Soil Association

NEW FROM FORDHALL

YOGTAIL PUNCH

We are making three new exciting foods, skilfully blended Live Fat Free Yoghourt, citrus fruit, and fresh berried fruits, Advocaat, Red Wine and Rum. These delicious Yogtails are a new approach to your party, slimming diet and mid-morning break, also make an excellent pick-me-up. They may be laced with extra wine, spirits, liqueurs or fresh fruit juices to make punch bowls or fruit salads, etc. A luxury sauce for puddings and sweets, add an egg and stir till thick over a pan of hot water.

ADVOCAAT YOGTAIL PUNCH

We make our own Advocaat on the farm, from free range eggs, Barbados sugar and brandy—skilfully blended with Fordhall Fat Free Yoghourt, and whole citrus fruits including tangerines, oranges and lemons, making this punch an excellent tonic for children of all ages.

RED WINE PUNCH

It is a unique blend of Armadillo wine, red currants, blackberries, damsons and fat free yoghourt. It has a clean, sharp taste, the perfect companion with your evening meal or late night snack. Laced with a favourite wine it makes a perfect Punch Bowl for cocktail parties. Also, for a particular occasion, set in a fancy jelly mould, better still using wine instead of water to dilute.

YOGTAIL PUNCH WITH RUM

This needs no explanation or description out of this world. Serve hot lived! We have added some berries just to increase yo

LIVE SKIM MILK
Yogtail Regd
WITH
BLACKCURRANTS & RUM
FROM
FORDHALL
5 FL. OZ.

The benefits of Bacteria

Testing products in the Fordhall laboratory

FORDHALL LIVE YOGHOURT CULTURE

ENQUIRE HERE

The benefits of bacteria: Why did yoghourt become a Fordhall passion?

The Oxford English Dictionary defines yoghourt as "a *slightly sour*, *thick liquid* made from *milk* with *bacteria added* to it, sometimes *eaten plain* and sometimes with *sugar*, *fruit*, etc. *added*."

Yoghourt is made by fermenting milk with the help of particular lactose-loving bacteria. These bacteria consume lactose – the sugar in the milk – and excrete lactic acid as a waste product, which is responsible for both the tart flavour and the thickening by acting on the casein protein in the milk.

However, twenty-first century yoghourt (in the main) is not sold as a live product. Instead it is sterilised in order to preserve it, thereby killing the beneficial bacteria that has been added to create it.

Yoghourt at Fordhall, like that of today, was produced using a culture of *Lactobacillus bulgaricus* bacteria (a Bulgarian strand of bacteria) and *Streptococcus thermophilus*; at the time they were imported from Europe.

MAKE YOUR OWN YOGHOURT...
        with "FORDHALL" Fresh Live Culture!

A 1 oz jar of culture will make delicious fresh yoghourt daily, or less often, for two or more weeks and costs only 5s post free. No special equipment is necessary apart from a cooking thermometer.

Our culture is made from fresh "Fordhall" Jersey or goats' milk and pure lactose culture *bacillus bulgaricus*. The medical profession are recommending it widely as a valuable aid to good health and for the treatment of diabetes, nerves, acidity, ulcers, constipation and for slimming, as well as for reconditioning the system after treatment with antibiotic drugs—which are notorious for killing off the useful as well as the unwanted bacteria in our bodies.

Send your name and address with a *crossed cheque* or *crossed postal order* for 5s to:

**Home and Cookery Centre,
60 New Oxford Street,
London, W.C.1.**

Fordhall Farm *The Yoghourt Years* 23

> "The bacteria cultures came in little vials from Switzerland by post; we stored them straight away in what I think was liquid nitrogen until we needed them. We propagated enough each day to provide sufficient amounts to culture a whole vat of milk... This was then added to a vat of milk which had been separated [the cream taken off to leave semi skimmed milk]. The vats were surrounded with hot water to gently heat the milk to the required temperature; after the culture was added, the vat had to remain sterile and remain incubated at a constant warm temperature for a number of hours. When it had set you had to cool it very quickly because you didn't want it to over grow, otherwise it would go too acid and this would not taste nice."
> **Former dairy worker**

Whilst yoghourt was something new to the palette of the British housewife, it came naturally to Arthur and May. The philosophy of encouraging "life and bacteria" to continue in their natural cycles in a way that was also beneficial to the human gut complemented their holistic view of farming and living.

Arthur's daughter Marianne briefly summarises the importance of bacteria:

> *"The point that Dad had was that it's the bacteria in the soil that is the secret to fertility. If the soil's not right, life isn't right. I could listen to Dad talk about that for ages. That was the point. If you dig deep and turn the soil right over with the conventional plough, you expose the soil to the sun. The sun sterilises the soil killing all the bacteria, and you are left with the bare minerals. You're not left with what the plant really needs for growth. So Dad's whole belief was that you should have this system of decaying organic matter... let it all decay and then you've got the perfect base for bacteria to grow in the soil, and when bacteria thrive in the soil that helps the plants – that was the basis of Dad's farming philosophy. The business about bacteria interestingly follows into yoghourt as well ... Now of course this is well known, and when I listen to them talking about the different commercial yoghourt drinks that will help the bacteria in your intestines... we were saying that when we were selling our yoghourt a long time ago, talking about the fact that our yoghourt was live, that the culture used to make it was live. It was all based on the idea that bacteria matter so much."*

Yoghourt was not as naturally accepted by the general population in the 1950s and 1960s as it is today.

"You invariably got, 'Live? LIVE! Well I'm not having anything live inside me!' And, 'Bacteria! I spend pounds washing away bacteria!' ...and all of that. Again, it was new, it was brand new" says staff member Terry Healey, who remembers selling the Fordhall yoghourts on market stalls in the 1960s.

Arthur recalls the commitment that May had to their own milk and its use for yoghourt in his book **The Farmer, the Plough and the Devil.**

> "It reinforced [May's] belief in the wisdom of always trying to work with nature... that's why she preferred to work with milk that hadn't been processed by heat or chemical sterilisation... Milk in which teeming millions of organisms have established a natural balance which only changes gradually in an orderly and predictable way...whereas milk that has been sterilised, while a fertile medium for bacterial, yeast and fungus growth, possesses no natural inhibitors. If it became contaminated

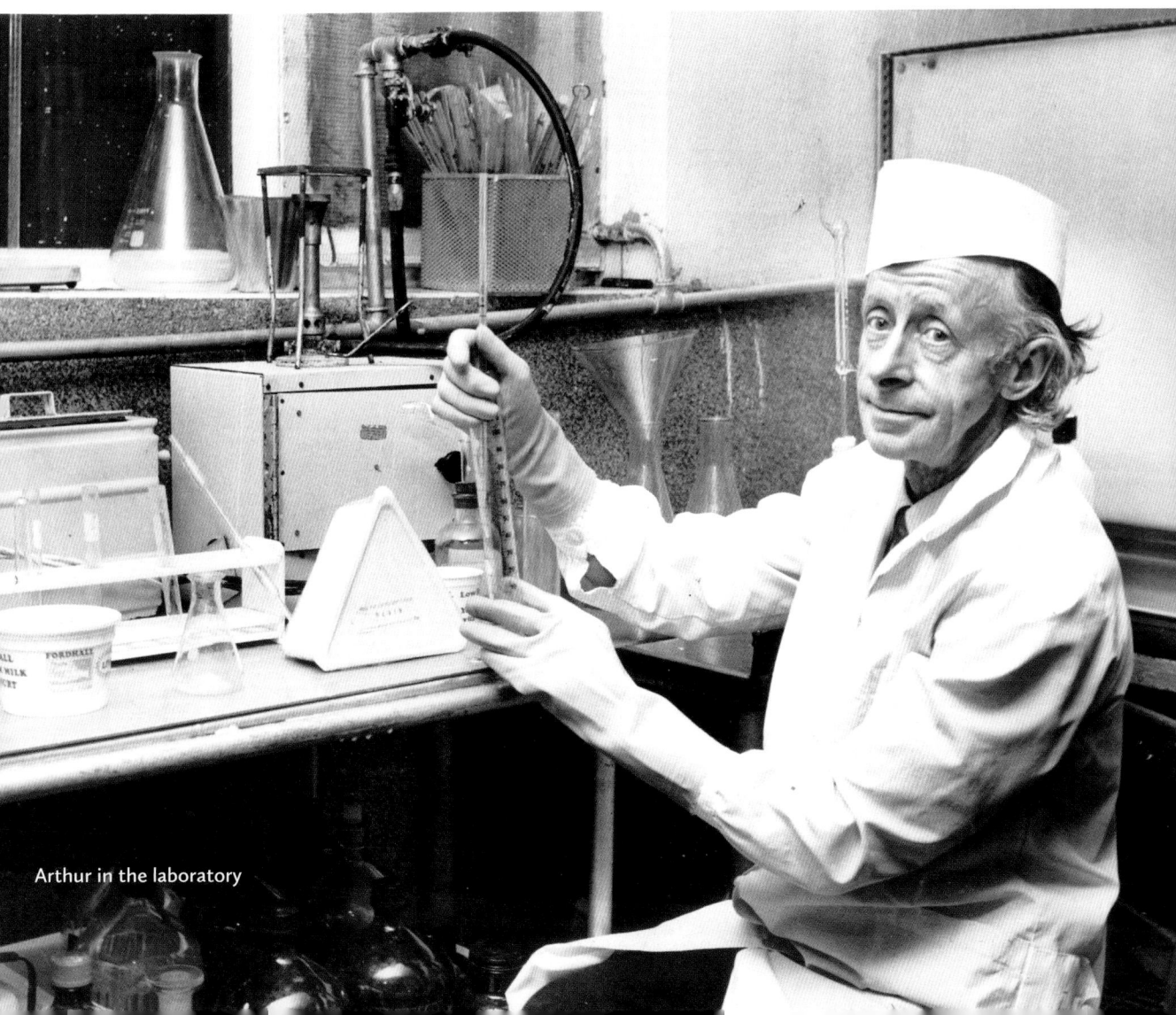

Arthur in the laboratory

with the wrong organisms, it wouldn't ferment to produce wholesome products; it would merely putrefy rapidly..."

May introduced yoghourt into Arthur's diet after he had problems digesting milk, cream and butter; he soon felt much fitter and slept better. With a sparkle in their eyes, a deep-rooted belief in the principles of yoghourt production and personal proof of its health benefits, their journey into commercial yoghourt production at Fordhall began. By the late 1960s, Arthur and May were even selling their own cultured bacteria for housewives to use at home.

MAKE YOUR OWN YOGHOURT!

We can send you fresh live yoghourt culture to make delicious health-giving yoghourt at home quickly and cheaply. A 1-oz. jar of our culture will make 3 pints of yoghourt, and by saving a jar from this yoghourt you can continue making yoghourt daily for from 1–2 weeks, depending upon the care you exercise in sterilising the jar each time to keep the culture pure.

Fordhall culture is made from fresh Fordhall Jersey or goats' milk and pure lactose culture, *bacillus bulgaricus*. Yoghourt made from it provides a valuable source of calcium and vitamin B and it corrects the balance of bacteria in the stomach, helps the glands to work naturally and, because it is easily digested, it is excellent for young children as well as for invalids. Babies can digest milk but adults, who find more difficulty in absorbing its full nourishment, find it more digestible and nourishing in the form of yoghourt. It is particularly valuable for invalids and elderly people with weak digestions.

In Bulgaria, hardly a meal goes by without yoghourt being served, and the people's general health and complexions are remarkably good. In America, and on the Continent, yoghourt is ordered and delivered daily with the milk. We in Britain are only just realising its great health value and doctors are recommending it widely for diabetes, nerves, acidity, ulcers, constipation, slimming and for reconditioning the system after treatment with antibiotic drugs. Antibiotics are notorious for killing off useful as well as unwanted bacteria and they can result in distressing side-effects. Yoghourt helps to replace natural, necessary bacteria in the intestines and wards off other ailments that attack us when we are "run down".

Preserved yoghourt, or yoghourt with artificial (chemical) "starters", lacks the culture so necessary if the full health benefit is to be derived from yoghourt. By protecting the stomach the live culture enables the body to function actively into old age and in this way its contribution to a longer, healthier life has earned it fame in Bulgaria.

Fordhall Farm, with its herd of Jerseys, has been organically farmed for over 29 years, and this gives added vitality to the milk used to make this live culture.

Some history:

By most accounts yoghourt was first created by Central Asian people in the Neolithic period. Analysis of the bacteria indicates that it may have originated on the surface of a plant. Milk may have become unintentionally infected through contact with plants, or bacteria may have been transferred via the udders of domestic milk-producing animals. In the early 1900s yoghourt production had reached the Americas and Europe.

It did not really become mainstream in the UK until the late 1960s to 1970s when large multinationals such as Ski and Eden Vale came onto the market, with additives, a longer shelf life and big advertising budgets. This was several years after the innovative May and Arthur Hollins brought it to the shelves of many a whole food store and specialist outlet.

Getting the product to market

Arthur with a cheese wheel

Getting the product to market

Fordhall's earliest outlet for cheese (between the wars) was the historic weekly market in nearby Market Drayton. In the mid-1950s, once rationing had ended and the Jersey herd providing quality organic milk and cream was established, Arthur and May's growing variety of dairy products could be distributed further afield.

After an encouraging start with Wolverhampton market, by the late 1950s Arthur and May were running market stalls in a number of midland and north-western towns, including Wellington, Shrewsbury, Chester, Birkenhead and Stockport. A wide range of products was also being supplied to food stores and delicatessens within a 70-mile radius, including Liverpool.

By the 1960s, thanks to successful stints at trade exhibitions including the 1962 London Olympia Food Fair, the northern

Two boys with an early Fordhall advertising board

May Hollins' market stall stocked with 1950s Fordhall produce

Lewis's chain and prestigious London stores such as Harrods and Selfridges were placing regular orders. Chris Clowes remembers Arthur's London trips: *"Twice a week, Helen who worked in the kitchen, washed Arthur a beige, linen-type jacket and he'd go off to Harrods with these deliveries."*

Invoices and record books stored in Shropshire Archives show that in 1973 Fordhall was selling over £1,000 worth of stock a month to nine Lewis's stores, including Selfridges in London. In today's money that would equate to a staggering £10,000 per month. Documents also show that they supplied Holland and Barrett stores from Lancashire to Surrey, Booths supermarket in Kendal, David Evans in Cardiff and the Health Food Stores chain in Edinburgh, Dumfries and Ludlow, among countless independent wholefood retailers the length and breadth of the country.

Fordhall Farm *The Yoghourt Years*

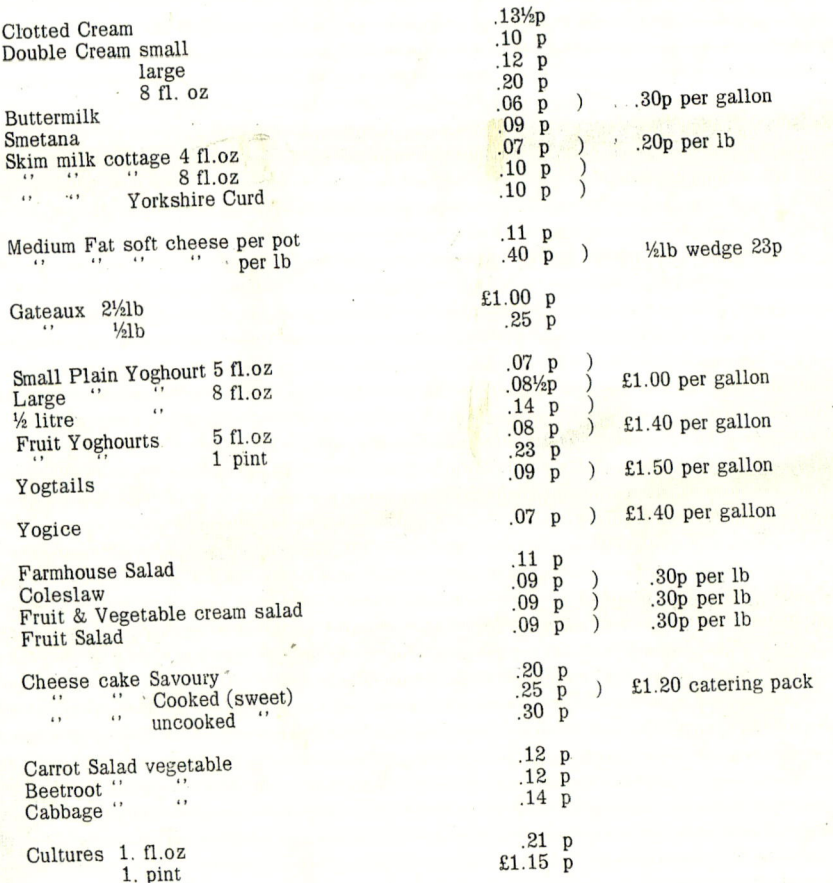

FORDHALL ORGANIC FARM

*Market Drayton
Salop
TF9 3PS*

Telephone Tern Hill 255/6

Specialities:
CLOTTED CREAM
SOFT CHEESES
YOGHOURTS
YOGTAILS
YOGICE
SALADS

Members of the Soil Association

Partners A. HOLLINS   I. M. HOLLINS

Prices as from May 13th 1974

Clotted Cream	.13½p	
Double Cream small	.10 p	
large	.12 p	
8 fl. oz	.20 p	
Buttermilk	.06 p)	.30p per gallon
Smetana	.09 p	
Skim milk cottage 4 fl.oz	.07 p)	.20p per lb
" " " 8 fl.oz	.10 p)	
" " Yorkshire Curd	.10 p)	
Medium Fat soft cheese per pot	.11 p	
" " " " " per lb	.40 p)	½lb wedge 23p
Gateaux 2½lb	£1.00 p	
" ½lb	.25 p	
Small Plain Yoghourt 5 fl.oz	.07 p)	
Large " " 8 fl.oz	.08½p)	£1.00 per gallon
½ litre " "	.14 p)	
Fruit Yoghourts 5 fl.oz	.08 p)	£1.40 per gallon
" " 1 pint	.23 p	
Yogtails	.09 p)	£1.50 per gallon
Yogice	.07 p)	£1.40 per gallon
Farmhouse Salad	.11 p	
Coleslaw	.09 p)	.30p per lb
Fruit & Vegetable cream salad	.09 p)	.30p per lb
Fruit Salad	.09 p)	.30p per lb
Cheese cake Savoury	.20 p	
" " Cooked (sweet)	.25 p)	£1.20 catering pack
" " uncooked "	.30 p	
Carrot Salad vegetable	.12 p	
Beetroot " "	.12 p	
Cabbage " "	.14 p	
Cultures 1. fl.oz	.21 p	
1. pint	£1.15 p	

Price list from 1974

"I lived in the farmhouse... I had two winters here; it was a cold house in 62 and 63. I remember at the time the Royal Canadian Air Force was at Tern Hill, in those days it was helicopter training... I know that Buntingsdale was then the district or regional command because we supplied the officers' mess there with cream, yoghourt, cheeses and everything. They wanted the best and they bought the best... On one occasion Arthur came into the cow shed at about half past five and said 'Tidy yourself up you're going to Birmingham'. Barrows of Birmingham were catering for an event at the Town Hall that evening and the Queen Mother was coming. They had had about a gallon and a half of cream, and they tipped it over and spilt it and they wanted another gallon and a half... and off I went to the middle of Brum with most haste, in that van (points to picture of white Commer van) with this small churn of cream. These were the exciting things that happened when you were working for a small company; you had to turn your hand to anything." **Mike Niccolls**

1960s Market stall

Fordhall Farm *The Yoghourt Years*

From pushbike to pantechnicon

In the early days, Arthur's overloaded tricycle proved an effective means of local transportation, as he recalled in an article from **The Listener** in 1971:

> "I bought a second-hand grocer's tricycle, and we packed clotted cream in tomato boxes which I stacked all-round the cycle and on my back. I must have looked a very funny sight to the stationmaster and to local farmers."

Distribution in the main was by road and rail, initially using a tractor and trailer to deliver goods to Market Drayton railway station. Local buses were also sometimes used, certainly to wholefood shops in Ludlow.

As the Fordhall Empire grew, more deliveries were made by road using liveried vans such as those shown.

In order to meet the growing demand for Fordhall produce, members of farm staff were often drafted in as delivery drivers, making increasingly long-haul journeys.

Although the roads were quieter in the 1960s and 1970s, vehicles were much less reliable than today, and the road network less well developed. Life as a delivery driver would have had its ups and downs, especially bearing in mind the long hours behind the wheel deemed acceptable in those days before regulation. Mary Cowen recalls a particularly busy period along with one of the perks of driving a Fordhall van:

> "On Monday, Wednesday and Friday, I would go to Wolverhampton, Birmingham, and Coventry. On Tuesday and Thursday I did Manchester, Liverpool and Southport, which was nice because I would spend the last couple of hours on the beach at Southport."

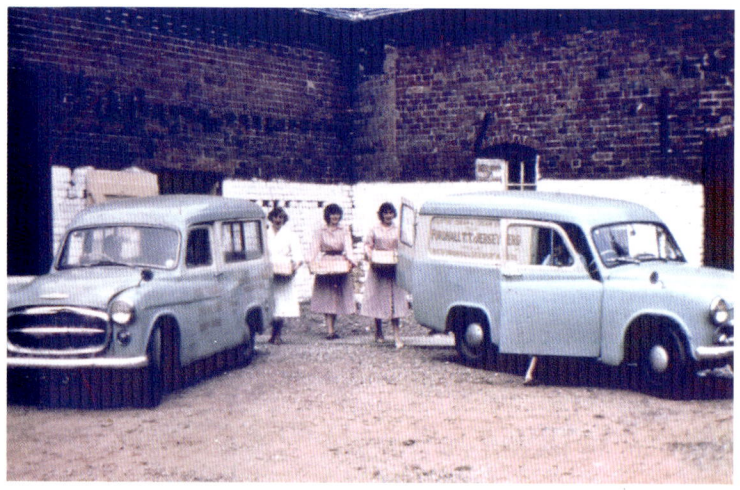

Top: 1950s Commer delivery vans
Above: 1960s Commer delivery van

Fordhall Farm *The Yoghourt Years*

Shop display with Fordhall van parked outside

> "After taking the orders we would put them up for the next day. The trays would come in from Fordhall; it was a tray of 20, so if they wanted so many of rum and coffee and so many strawberry then we would make up a mixed tray, or one tray of strawberry and one tray of raspberry."
>
> **Pat Davis, wife of Eric Davis**

Due to the sheer scale of the operation, Arthur was inevitably obliged to explore other options for distribution. He employed sales agents for each region of the country, who would meet with local stores, demonstrate products and arrange deliveries from nearby railway stations to enable Fordhall to cover every corner of the country. But using public transport, with its lack of refrigeration, was also problematic, so Arthur and May later employed distribution agents. Eric Davis remembers his time as Fordhall's distribution agent after 1969:

> *"When I first made contact with Arthur Hollins he had already forged links with some of the major department stores: there was Lewis's in Hanley in the North, Dingles at Plymouth in the South, other stores in Gloucester, Cheltenham and Bristol, and in London, Selfridges. So these needed to be serviced on a regular basis. Arthur didn't have a reliable distribution setup, so basically we endeavoured using those key Fordhall outlets. I became an independent distribution agent for Fordhall and developed runs five days a week to service these places and then progressively contacted all the health food stores and delicatessens on the route to get Fordhall products in those depots, in those shops and stores."*

1960s Stocking shelves with Fordhall products

Finally, distribution was farmed out to a firm called Simons Enterprises, based in Spitalfields in London. Simons had cold stores in Titchfield in Hampshire, Pershore in Worcestershire and Ely in Cambridgeshire, and employed their own delivery drivers.

At its peak, Fordhall was supplying over 100 outlets across the UK. In the early 1970s, Arthur reportedly made a move to sell yoghourt to the French, and he even investigated establishing licensing arrangements for Fordhall products as far afield as America and Australia. Although we know a licence was made to the monks of Caldey Island off the Pembrokeshire coast in the mid-1970s (for a 5% commission fee), evidence suggests that nothing concrete ever came of these more international ambitions.

"I remember going to the Ideal Home Exhibition and being there for a couple of days, or even three days or four days; selling Fordhall yoghourt, which was extremely nice. It's the best yoghourt you have ever had. They used to do cheese, and wine yoghourts. Arthur was into packaged carton salads, yoghourt milk, and juices, long before anybody else. But one of the problems, I think was the fact that he was before his time. It was only about 10 years later when people were actually putting it in their shopping basket and it became a good thing." *Jill Blud*

"'We've got this chap who's making this amazing yoghourt, and it's frozen and it's organic and he's very fussy and he's very bossy. We need it delivered to Selfridges and places.' Me and my little van did it for three pounds an hour. We picked it up from a huge cold store in Stamford Street-on the corner of Duchy Street in London. All the yoghourts were in there, frozen, ready to take out... I took them to Selfridges and other places – I can't really remember – but there was... a South West Guildford and Woking Store... I think they were delis, and rather smart grocery shops." *Pam Price*

Building the brand

FORDHALL FARM FOODS
PURE · LIVE · NUTRITIOUS

1966 Arthur's son Robert, demonstrating at a Manchester store

Building the brand

Although Fordhall started out as a small local enterprise, the business grew at an astonishing rate thanks to the tireless efforts of Arthur and May. May's niece Mary Cowen recalls one of Arthur's quirky methods for enlarging the market for Fordhall produce in the early days:

> "We would get the bus at the bottom of the drive into Shrewsbury, get whatever train that happened to be there, whatever took our fancy and we would go off somewhere. When we got to that town, Arthur would enquire of people around, 'Which is your highest class food shop here?' and off we would go. He would have done very well on 'The Apprentice'. He would pitch his wares to the buyer and every single time would come away with orders, but he stuck to the principle of one shop per town."

Above: Late 1960s Food fair Belle Vue, Manchester
Below: 1973 Display fridge

Fordhall Farm *The Yoghourt Years*

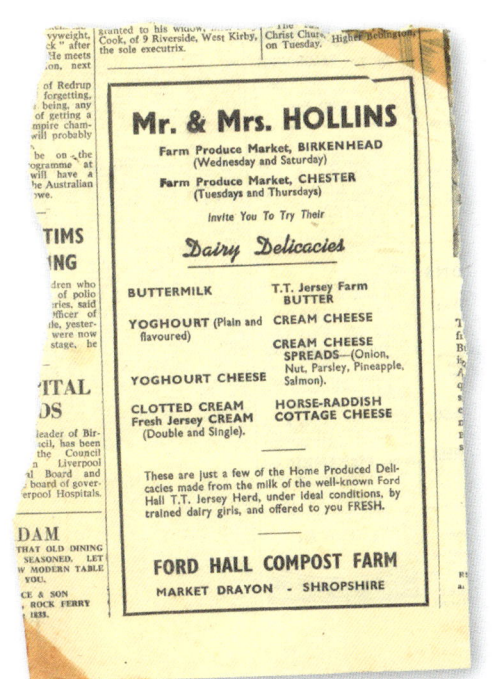

As Fordhall's products were 'natural' (what we would now call organic) and 'different', customers needed to be convinced of their health-giving qualities before deciding to buy. Arthur and May were skilled at demonstrating their wares, and successfully built up the brand. As more department stores around the country began to stock Fordhall products, full-time demonstrators were recruited locally to spread the word. Terry Healey, who worked for Fordhall in the 1960s, remembers life as a demonstrator from his days in Lewis's department store in the Potteries:

"You would arrive with your little white coat on and the cheeses were in large triangles, plastic triangles, not in little boxes, with a cellophane top on them. You used to take the top off, you had a little tub and if they wanted four ounces you put four ounces in a tub or two ounces and weigh it. And if they wanted a taste we had little plastic spoons and we used to give them a taste. Same with yoghourts, you had to do that as well."

Terry went on to describe some of the challenges involved:

"You had a problem inasmuch as it was cottage cheese and whereas today we accept cottage cheese as part of life, it wasn't in the 60s, and it was curd and if you start talking about soured milk, and curd... people in the 60s looked at you odd! So it wasn't as easy as you would think nowadays to say, 'Well, try a bit of this cottage cheese with chives or tomato or something,' and you'd always get someone saying, 'It's off.'"

However big the operation became, Arthur was careful not to lose sight of his customers, as he explained to **The Listener** on 2 December 1971:

"Our present marketing ideas have developed from direct contact with our customers, and if we maintain that close link, we feel we will always hold onto our small portion of the market, which grows slowly on trust and understanding, and a feeling that what we offer contributes to the health and fitness of the whole family. This creates a long customer life, a link that must not be broken. The same link exists between our plants and their soil friends."

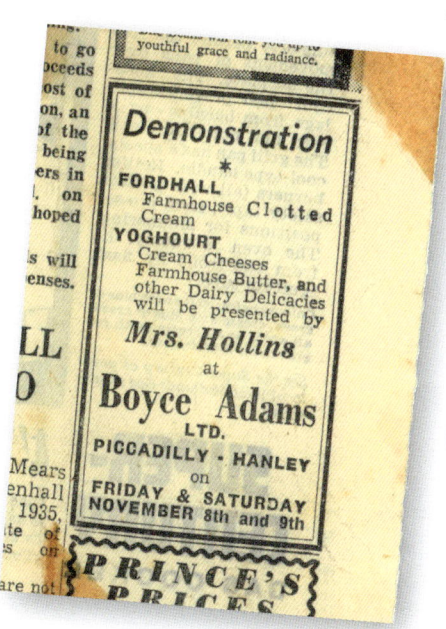

1950s newspaper cuttings

Fordhall Farm The Yoghourt Years

Family and friends all helped to promote the Fordhall range

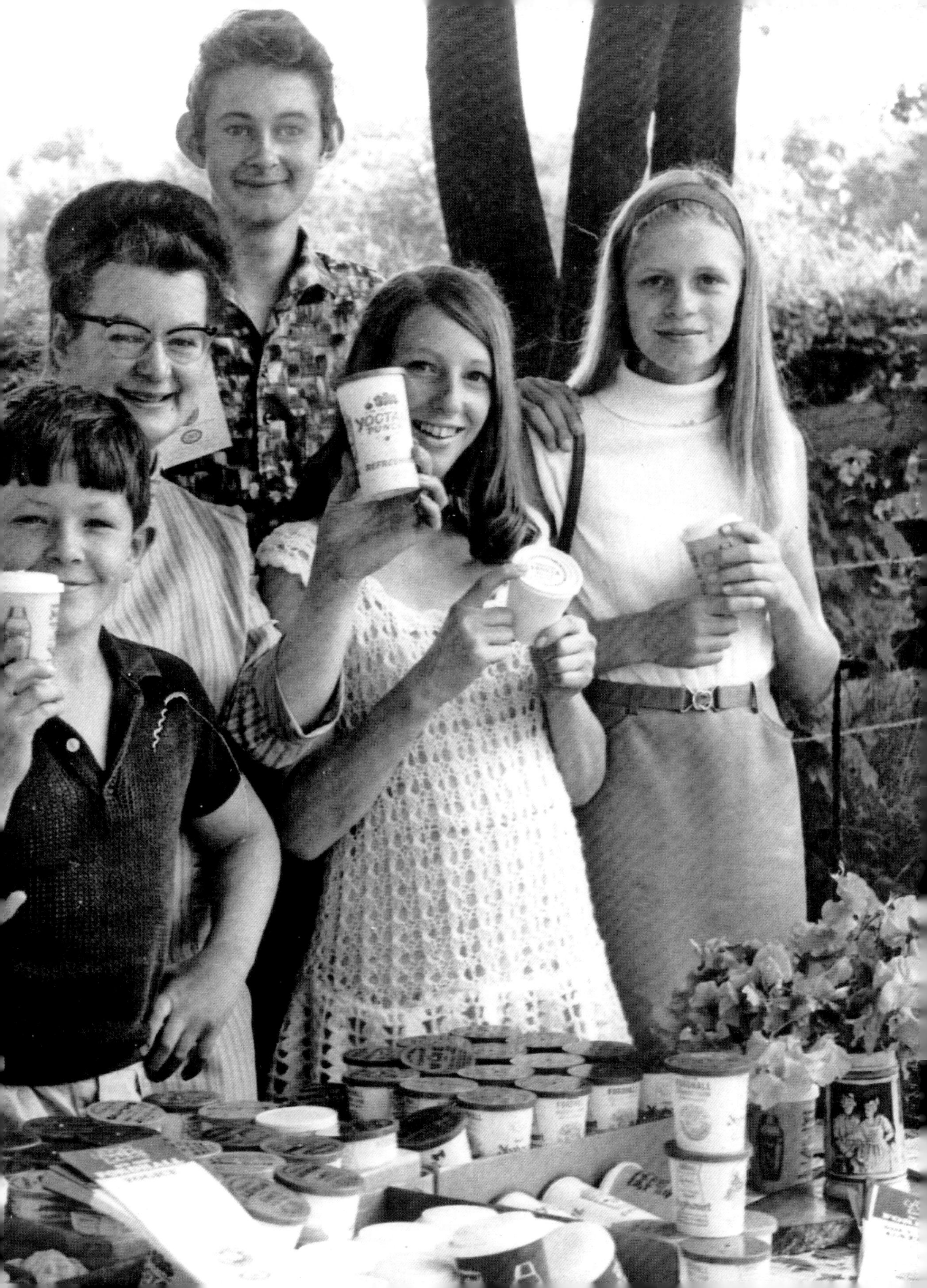

Left: 1950s Liverpool
Above: May at Jenkinsons, Stafford
Below: 1959 Manchester Food Fair

Above: 1961 London Food Fair
Right: 1964 John Williamson and May Hollins exhibiting at Belle Vue Manchester

Fordhall Farm *The Yoghourt Years* 41

Developing the dairy

Robert Hollins in the milking parlour

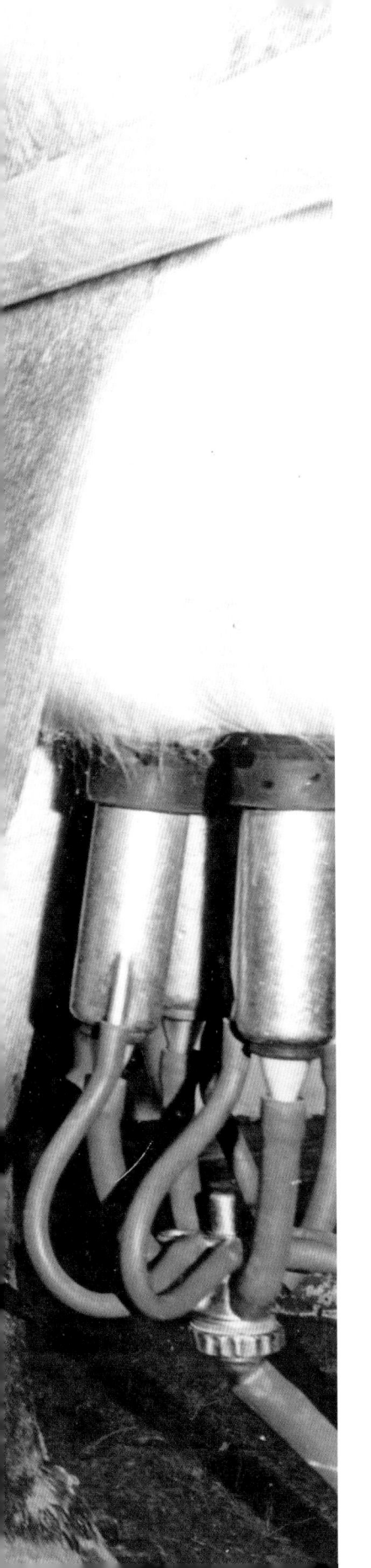

Developing the dairy – from cowshed to factory

In the early days of Arthur's tenure at Fordhall, the dairying operation was on a small scale. Mary Cowen remembers hand-milking at Fordhall as a girl in 1942, but by the time she returned in the early 1950s, "he had a herd of Jersey cows and a lovely new shippon (cowshed) with 32 stalls and a (portable) milking machine."

Due to the popularity of their cream cheese, yoghourt and cottage cheese, Arthur soon decided that the new shippon would become the new 'factory'. So the Jerseys, his pride and joy, were moved to a modern purpose-built Fullwood and Brand herringbone milking parlour built in 1962-1963. This would have been the envy of neighbouring farmers at the time, as it was one of the first installed in the area. Everything was arranged to be as efficient as possible, with the milk coming from the cows in the milking parlour next door to the new factory through sparkling clean pipes and separated into large heated vats for cream, yoghourt and cheese production.

1961 Deep freeze unit

Fordhall Farm *The Yoghourt Years*   43

Above: The shippon is upgraded to allow more cattle to be milked and a modern portable milking machine is installed
Right: An article from the Newport Advertiser from March 1965 showing Arthur's new milking parlour
Below: When the new milking parlour was built in 1963 the shippon became a gleaming new factory

44   **Fordhall Farm** *The Yoghourt Years*

"I used the machine that made the triangular cartons for the soft cheese when I had spare time. It arrived whilst I was working at Fordhall (in 1963) – you used to have to switch it on, wait for it to heat up and when it reached the right temperature you pressed the lever and it made two of the triangular cartons. You had to cut it with a knife in between and then stack them so they were ready to be filled. Then, when they were filled, you put a film of cellophane or plastic over the top. I think it was heat sealed." **Jean McAdie**

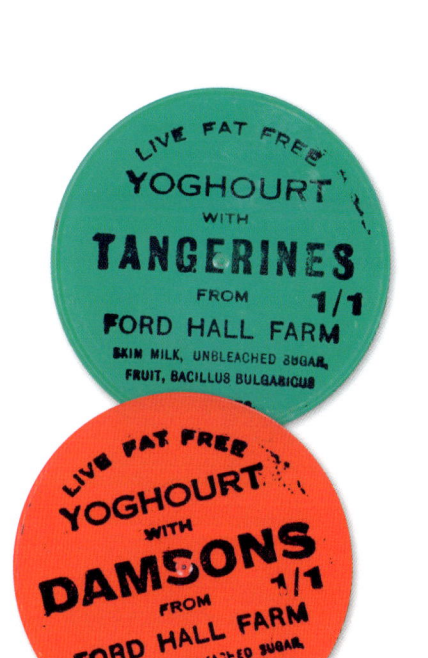

Change was in the air, and as the operation grew, so did the variety of processes that had to be carried out. Arthur was highly inventive and considerably ahead of other small businesses. In 1961, to aid efficiency with batch production and to extend shelf life, he installed a large deep-freeze unit which held supplies for local markets, and he later converted the old bull pen into a large walk-in deep-freeze.

In his habitually unorthodox way, Arthur altered and fixed machinery to meet his needs. This included adapting a stainless steel paint-filling machine to dispense yoghourt automatically; to mix ingredients into the cheeses he used a baker's dough mixer, then portioned it into containers with a sausage-filling machine.

The soft cheese was packed in large triangles and heat-sealed. Although most pots were bought in, the plastic triangles were made on site with their own moulding machine, giving the enterprise full flexibility. Eight triangles fitted into a circular revolving serving tray designed by Arthur, ready for selling loose to the customer.

In the packing department, machines filled specially designed plastic yoghourt pots printed with the farm logo. According to Terry Healey: *"We used to print our own lids. We had a 'John Bull' printing set and there were three or four different colours, and you had a different stencil for each flavour, so if you were doing a raspberry or a strawberry you'd use a red lid. Then you'd put this little plastic lid on the machine, pull the handle down and it would stamp "Fordhall Farm" on*

"The lids all had different colours for the different flavours. … we had quite a bit of trouble with the printing machine at first, it would miss bits and things and Arthur was always tinkering around to get it nice and clear… The machine printed on plastic as well as paper. It was an adaptation; I think it had to have padding to take the plastic lids. Arthur was very keen on being sort of, self-sufficient; everything was done here as much as possible." **Jean McAdie**

Fordhall Farm *The Yoghourt Years*   45

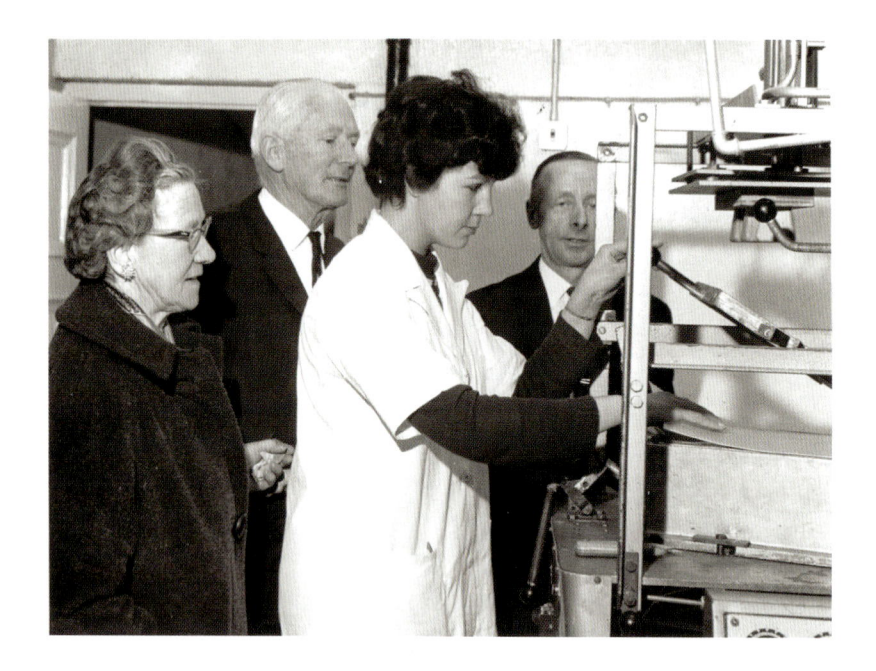

Above: 1960s Yoghourt pots are filled and lids fitted before packing for distribution
Left: 1965 Jean McAdie making triangular plastic containers for the soft cheese
Below: 1973 Display fridge

it. So that was another job you used to have to do every day, stamp up the lids for that days "make". Arthur, of course, had adapted a paper printing machine to print on the plastic pot lids.

Mixing, packaging and cold storage on the farm were not the only challenges facing Arthur and May; in delivering further and further afield, they also had the challenge of ensuring their 'live' products remained fresh and attractive in shops and on market stalls. As nothing suitable was manufactured at that time, Arthur was continually inventing cold storage solutions, including a display refrigerator in partnership with Simons Enterprises Ltd, which made the promotion of Fordhall products easier for retailers and self-service possible for customers. This allowed the whole range of Fordhall products to have maximum exposure in stores across the country. Never one to let the grass grow under his feet, at one time Arthur even considered developing automatic cold milk vending machines for his market stalls, but decided that this was not viable and abandoned the idea. Ironically, this method of selling 'raw milk' is promoted as a novelty in Selfridges today.

By the end of the 1960s the demand for Fordhall products had grown exponentially. Arthur and May were struggling to keep up with the success of their extensive product range and they were outgrowing their new factory which had previously taken over the new shippon. Wanting to expand, but keep production on site, in 1969 they looked to build a 3,000 cubic foot factory at Fordhall by purchasing an old pumping station. However, this was not to be. Fortunately, Arthur's method of outdoor grazing for the cattle

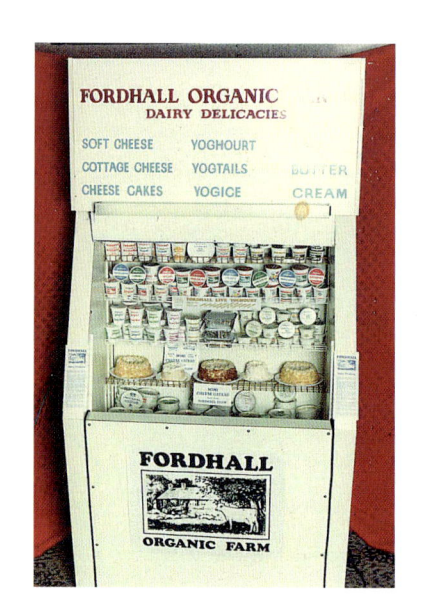

46   Fordhall Farm *The Yoghourt Years*

FORDHALL ORGANIC FARM

Savoury Cheese Cake
(for home cooking)

Ingredients: Med. Fat Soft Cheese, Onions and other veg. in season, Wholemeal Flour, Egg, Spices and Salt.

Cooking Instructions:
Remove lid and cook when thawed in moderate oven (350°F or Mark 4 gas) for 1–1½ hrs.
EAT HOT OR COLD

Fordhall Organic Farm Market Drayton Shropshire

Right: Hard at work in the factory
Below: As the enterprise advanced so did their equipment. The dairy moved away from hand packing to conveyor belts and speed bagging

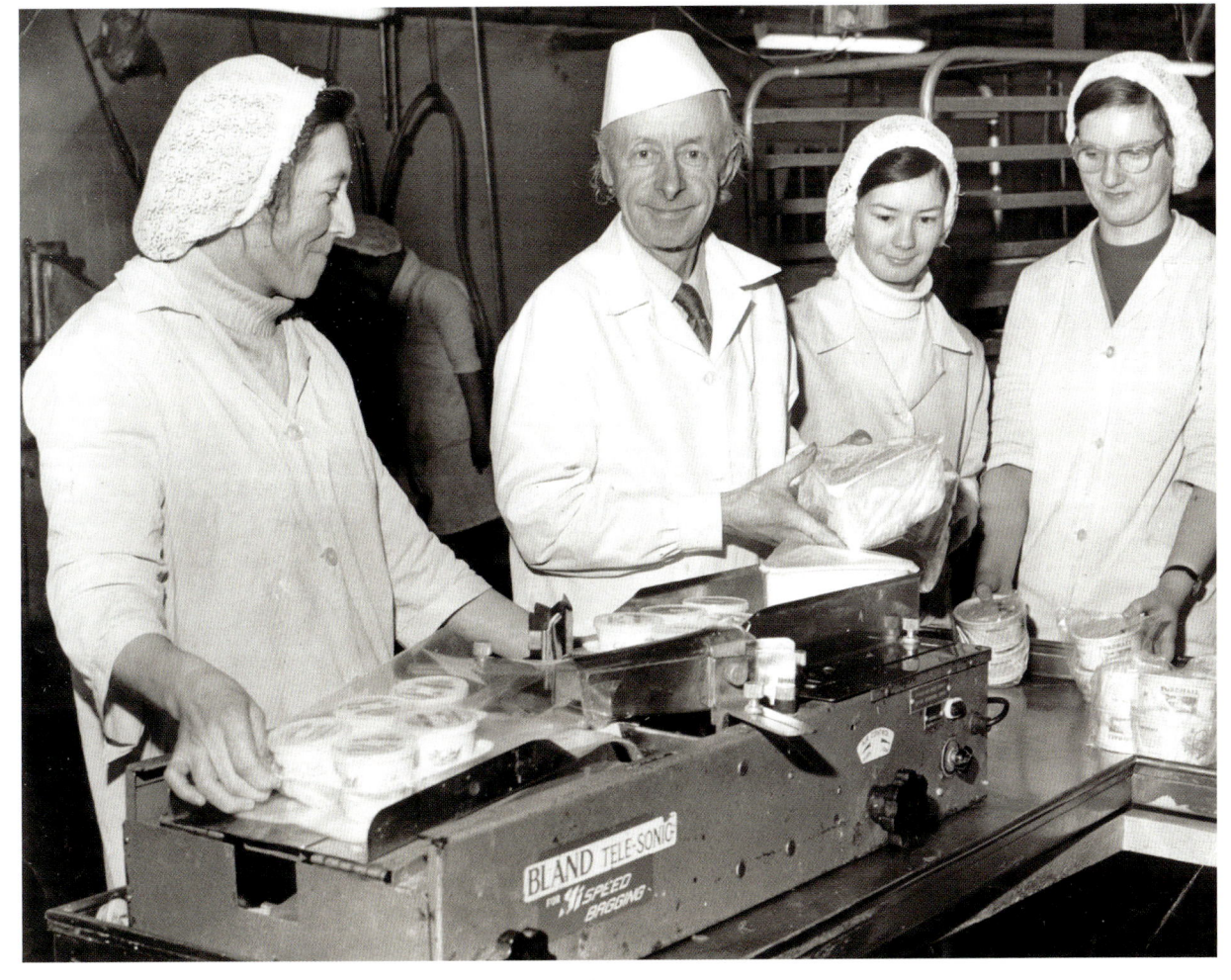

Fordhall Farm *The Yoghourt Years* 47

1963 Arthur selling triangular soft cheeses in Debenhams, Romford, using a revolving tray with his innovative hanging refrigeration system.

ATTRACTIVE EMPLOYMENT FOR AUSTRALIANS AND NEW ZEALANDERS

DAIRY MAIDS required for cream production, cheese making, yoghourt making. Previous experience not necessary if willing to be trained. Opportunities to demonstrate products throughout Great Britain. Caravan accommodation on the farm during Summer, Flats ready for the Winter. Minimum period of employment 2 to 4 months, or permanent with responsible position. Excellent post now available in London, with responsibility. Appointments can be made for interview in London. Write to: Hollins, Fordhall Farm, Market Drayton, Salop, Shropshire. Phone: Ternhill 255.

FORD HALL FARM, MARKET DRAYTON, SALOP.

had conveniently left more building space for their yoghourt empire to develop, so instead they added cladding and a floor to the Dutch barn (which had been the cattle's hay store), turning this into a new storage area. With more space in the existing factory they created additional walk-in fridges, production area and even extra toilet facilities for their growing workforce.

By now, Arthur and May were competing with the giants of the soft cheese and yoghourt industry; their farmhouse operation had grown so big that the milk from their 100 Jersey cows was no longer enough to supply their customers. To meet demand, extra milk was bought in from neighbouring Jersey herds which, like Fordhall's, were regulated by the Soil Association. They even established a small laboratory adjacent to the dairy for testing the products, in the charge of Gillian Davies, a trained bacteriologist.

Terry Jakeman worked in the dairy when he left school in the 1960s, and describes it as a hive of industry: *"Loads of women with their fancy hats on, and blue or blue and white overalls. I can remember the fluffy hats they used to wear. Cloth hats, I'm sure they were, like a nylon hat and everyone was working, and the sound of the machines, some making butter, and girls with the yoghourts, the yoghourt pots dropping, going under a machine filling them full of yoghourt, others collecting them at the other end. Everyone seemed to be working. It was quite noisy as well, with the machines cutting the vegetables up."*

In 1971, an article in the **Liverpool Daily Post** said the farm had *"a staff of 40 women in the modern dairy busy processing, packing and distributing a wide variety of dairy products...and in addition to the dairy staff, about 10 girls out giving demonstrations".* This had grown from a staff of six only a decade before.

"Arthur tried all sorts of enterprises to keep the place afloat. He was a man full of ideas and he would not be stopped in his attempt to try these ideas until it nearly came apart at the seams. Until then Arthur was hell bent that they were going to work, and some did work and some didn't. There was the replacement of the roof that was blown off by the wind, he managed to lever it back into position – most people would have taken it apart and had it rebuilt – we put two poles under it and gave it a good pull and it came back into position. Nobody would have believed it could have worked, no one expected it to work, but Arthur was intent on getting it to work – and it did!" **Mike Niccolls**

Above: 1960s Dairying processes
Left: 1970s Making cream cheese gateaux

50   Fordhall Farm *The Yoghourt Years*

Fun in the dairy

Arthur playing hide-and-seek

1964 Birdcage carnival float

Fun in the dairy

It is clear that the infrastructure at Fordhall evolved with the same creative energy and enthusiasm that accompanied everything Arthur and May touched. Terry Healey sums up the feeling: *"The more I thought about it the more I've thought, I really did like those people... I love enthusiastic people. They certainly helped me out when I left school. I learned enthusiasm from May. You couldn't not learn it, and from Arthur in his own way. Yes they were lovely people... and the people who worked there were lovely... it was a family, that's the way you could describe it, the Fordhall family."*

There was great camaraderie and the dairy staff enjoyed many lighter moments; birthday rituals, sing-songs, parties and even staff outings are all remembered with fondness.

Pete Guildford recalls: *"I was only around 18 when I first started and when I saw all the lovely girls who worked there I used to think I was in seventh heaven."* The atmosphere was happy; some of the younger workers thought a few of the older women were "naggy" but that didn't stop the fun. Pete's birthday coincided with that of one of the other men: *"I said to Colin, 'I can hear the girls plotting something and I think we are in for it.' As work for the day ended, we were both set upon by lots of the girls. They pinned us down... then out came the liquid chocolate and both myself and Colin were smothered in the stuff; our shirts were torn open and the girls covered us. I recall going home and my mother saying 'I bet you enjoyed that', well I must confess we both did."*

That wasn't the only birthday tradition. Terry Jakeman said: *"There were these big vats. You had to scrub them out all the time. But if it was your birthday, they would be filled with water and you went in it! The girls would pick you up and put you in. And vice versa: the three lads, they would pick up a girl and put her in it! One thing I do remember... it was one of the girls' birthdays and we got the vat and filled it with water. But what I did and no one did for any of us – I put a gas ring under it and warmed up the water. She knew she was going in at five o'clock at the end of the day but I was able to tell her that I'd got the water nice and warm for her."*

1960s Shotgun Wedding float in the Market Drayton Carnival

VISIT TO T.V. STUDIO

IF Mr. and Mrs. A. Hollins and their staff at Fordhall Farm, Market Drayton, wish to recapture some of the fun they had on the outing they took on Monday all they will have to do is turn on the television set in a week or two. There is a good chance that they will see themselves as the studio audience at the beginning of Granada's "Criss Cross Quiz".

A party of about 45 of them went to Granada's Manchester studio and watched the recording of the programme plus two episodes of "Coronation Street" and afterwards they met the stars from both popular shows.

The night didn't end there either for they went on to the Manchester United Football Club for a cocktail party. It was quite a home from home because what should be served up but savoury cheeses made back on the farm at Fordhall.

To close the outing the party visited the Embassy Club and saw singer Danny Williams.

1966 Fordhall visit to Granada Studios and Manchester United FC

"The one that sticks in my mind was working at the Ideal Home Exhibition. I've got a feeling that it was in Birmingham... But we were there, in our white coats as was expected. ...and I was with Mrs Hollins – May – and we ran out of stock. Now the rule at these exhibitions is that you can't stock up your shelves once the public are in. So Mrs Hollins used her initiative and saw a stretcher, because there were stretchers on the wall by the first aid spots... So she grabbed this stretcher and myself and her shot out with it, in our white coats. We loaded up off the van with some more cheese and yoghourt, covered it up with a blanket and came in, with Mrs Hollins leading the way saying "Excuse me. Excuse me. Emergency"... and we're carrying this stretcher in our white coats. Of course, people scattered in all directions to let us through. We got back to the stand and shot the stuff behind the counter and put the stretcher back up again!... And that was Mrs Hollins that was just her." **Terry Healey**

The Fordhall Song

For a time in the 1960s and 1970s, there was a vogue for writing and recording school and workplace songs, and Fordhall was no exception. Written by Arthur, here are the words to the Fordhall Dairymaids' song:

*We're from the Fordhall Farm
All dairy maids are we
We bring you cream and cheeses
Fresh as fresh can be
Cottage cheeses for your diet
Diabetics too
Thirty-two varieties we bring to you*

*Our creams are unpreserved,
They're pure and healthy too
True yoghourt is the milk
We offer now to you
Trifle in fruit and cream
And great variety
Cheese in salads
Or as tasty bits for tea*

*We're from the Fordhall Farm
All dairy maids are we
We bring you cream and cheeses
Fresh as fresh can be
Cottage cheeses for your diet
Diabetics too
Thirty-two varieties we bring to you
We're from the Fordhall Farm
All dairy maids are we*

Terry also played an important role in facilitating lunchtime bathing parties at the farm: *"When I was in the dairy, being the youngest one there, the girls used to get me to do everything. They used to get me to get the thermometer out of the milk, wash it, put it in the garden pool and take the temperature! If it was the right temperature, the girls would go swimming in there in the dinner hour. Then I would have to take it back to the dairy and wash it again!"*

Terry Healey said: *"The one thing I remember... was the singing. Because Barbara Hollins, she'd got a lovely voice... and there were the Holding girls... They'd got lovely voices and I think were members of the operatic society and you'd hear them singing away quite merrily as they were working. It was lovely."*

On the creative front, Fordhall Farm entered the Market Drayton carnival several times in the 1960s. One memorable "Shotgun Wedding" themed float was based on the popular TV series 'The Beverly Hillbillies', whilst the inspiration for "The Bird Cage" is lost in the mists of time.

Outside work, lively times were to be had wherever the young female workers lodged. Terry Healey said: *"The vast majority were local girls, but... they employed people from out of the area. And they had rooms in the house... Somebody else had a caravan as well. They had a cottage going towards Müller's... They owned those and a couple of girls used to live in those as well. So there were outsiders, but they were mainly local."* Pete Guildford said: *"Oh great parties were held in the two cottages that some of the girls shared, the swinging sixties I guess you could call it."*

The staff went on coach trips too. The **Newport Advertiser** of 1966 said: *"Arthur and his employees take a trip to Granada Studios in Manchester in July 1966 to view a recording of Criss-Cross-Quiz plus two episodes of Coronation Street. They met the stars afterwards. They continued to the home of Manchester United Football Club at Old Trafford where they had a cocktail party and they were served savoury snacks which consisted of the very same cheese made at Fordhall Farm."*

Chris Clowes remembers life at Fordhall with a smile...

"They had coach loads of visitors. Yes, I think they came to the house for afternoon tea. In February (May) used to go out and stick her plastic daffodils in the flower pots!"

Fordhall Farm *The Yoghourt Years*

A warm welcome at Fordhall

Arthur uses an impromptu podium to address a visiting group

A warm welcome at Fordhall

From the earliest days of their wholefood dairy operation, Arthur and May welcomed visitors and weekend guests to Fordhall Farm. There are dozens of photographs in the Fordhall archive which show parties of happy tourists from the 1950s through to the 1970s.

By the early 1960s Arthur and May had a thriving visitors' scene, inviting organised parties and customers to spend a day at the farm on summer weekends, sampling cream teas and relaxing in the farmhouse garden. For the more adventurous, there was the prospect of a dip in the outdoor swimming pool (built to provide an emergency wartime water supply for firefighting), a game of tennis, crazy golf, or even archery. Keen as they were to make customers welcome at Fordhall, Arthur and May must have been juggling an enormous workload between hospitality, demonstrating, managing staff and overseeing orders, deliveries and marketing.

By the early 1970s, thousands of visitors per year were being conducted round the farm in daily tours from March to October. This number reportedly peaked at 8,000 in 1974 according to an article in the **Farmers Guardian**, and the typical visitor could expect a dairy and factory tour and a cream tea with a lavish helping of Fordhall's organic philosophy. Ever the showman, Arthur delighted in conducting tours and giving talks on his unique approach to soil development on which the whole Fordhall project depended. By the late 1970s he was often to be found hosting large meals and even fondue parties at the farmhouse; his enthusiasm never faltered.

Party conveyed by Jones Coachways Ltd, Liverpool

Fordhall Farm *The Yoghourt Years* 57

Above left: 1950s Visitors with Arthur and a goat
Above right: June 1961 Grassendale Towns Womens Guild
Left: 1971 Stan Jones' Herdsman visit'
Below left: A sunny day for a visit to the farm
Below: 1970s Visiting nuns in the farmhouse garden

58   Fordhall Farm *The Yoghourt Years*

Above: 1960s Class visit with a Jersey cow
Below: Arthur stood proud after another group tour

Fordhall Farm *The Yoghourt Years*

Fordhall Organic Farm, Market Drayton, was the venue on Thursday last week for 25 members of St. Stephen's Church Ladies' Guild, Crewe. Mrs. Mary Smith, the guild's social secretary, organised the outing and during the day members saw how butter, cream and yoghurt were made.

September 1971, The Chronicle

"Mr Hollins came up with the idea of inviting coach trips to look around the farm and the dairy. So these elderly ladies, usually from the W.I.s and nearly always ladies, used to turn up on the coach, and Arthur would meet them in his wellies… Eventually I took that role over. But he'd take them in to the milking parlour – and we knew what went on the milking parlour! As soon as the cow did it, you got the hose pipe and washed it away. But invariably, Arthur would walk in with a crowd of ladies and of course one of the cows would let rip all over the place. And of course these people, they knew a cow, they'd seen it on television, but they didn't know they did things like that! And pee! Well it's like a torrent when they get started. They were just shocked… the idea that this is good healthy food – 'Ooh, there's poo all over the place!'" **Terry Healey**.

1964 The cattle in the new herringbone milking parlour could catch you out

60   Fordhall Farm *The Yoghourt Years*

Arthur giving a talk in the 1970s

Visit 'FORDHALL WAY' and relax,

in the tranquility of the Shropshire countryside:

First class vegetarian and wholefood guest house with delightfully varied menu. Feel fitter for well balanced food and relish the flavour of organic produce.

FORDHALL WAY
PELL WALL COURT

A country house with charm and character delightfully nestled amidst acres of fine woodlands with many interesting walks, and within easy reach of Market Drayton - a peaceful country town on the Shropshire Union Canal.

Although it is an ideal centre for walking or touring, visitors must not miss the famous Wednesday street market with its many bargains.

Although Arthur played a pivotal role in Fordhall's success, May's contribution was equally important. It was her vision that underpinned a unique enterprise on the other side of Market Drayton which, but for a tragic accident, might have been the start of a healthy living movement. Fordhall Way, as it became known, was a vegetarian and wholefood guest house, established at Pell Wall Court in 1972. Guests were invited to embrace vegetarian eating and exercise, thereby living 'the Fordhall way' for a few days in a spacious country house setting, before returning home revitalised.

People with all kinds of dietary and exercise needs were catered for at Fordhall Way, and Fordhall products were much in evidence, as can be seen from this description of a 1974 wedding breakfast from the **North Shropshire Echo**:

> "The menu for the breakfast consisted of a green salad with two savouries of Fordhall cheese with onion and soya, nut and sweet corn. It was followed by a ribbon salad together with Fordhall fruit and vegetable salad, dates, nuts and raisins. These were complemented by pimento dressing and whole-brown bread rolls. Sweets varied from Fordhall Fruit Salad laced with Cointreau, Fordhall Cherry Cheesecake to a selection of Fordhall Gateaux and biscuits followed by dandelion coffee with honey."

1980s Arthur with his meat products at Newcastle-under-Lyme indoor market

End of an era

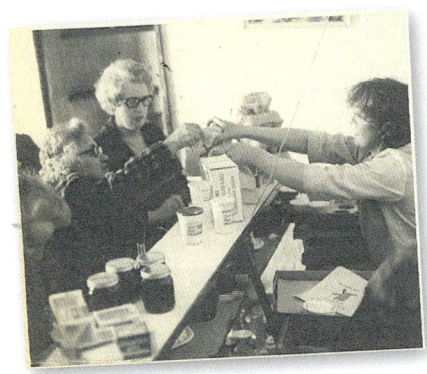

1972 Onsite farm shop

Arthur and his book 'The Farmer, The Plough and the Devil'

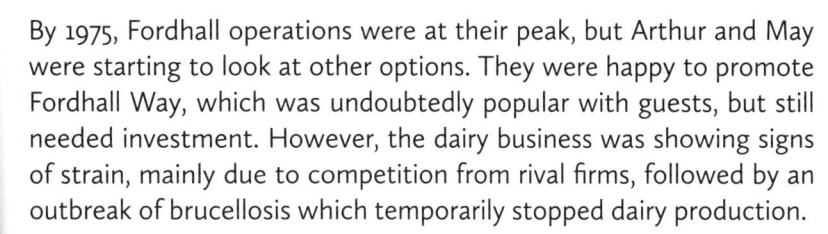

By 1975, Fordhall operations were at their peak, but Arthur and May were starting to look at other options. They were happy to promote Fordhall Way, which was undoubtedly popular with guests, but still needed investment. However, the dairy business was showing signs of strain, mainly due to competition from rival firms, followed by an outbreak of brucellosis which temporarily stopped dairy production.

Ski Yoghurt was launched in the UK by a Swiss company in 1963, and the market grew quickly. Other rivals such as Eden Vale and St Ivel soon appeared, and supermarkets developed own-brand yoghourts which could be offered more cheaply than branded products. These new commercial manufacturers had money to invest in advertising, and were not afraid to use sugar and other additives to lengthen the shelf life of their products, unlike Fordhall's which were natural and unadulterated. Fordhall's products had always commanded premium prices in high-quality outlets such as department stores, delicatessens and health food shops. But faced with competition from their cheaper, slickly packaged, highly sugared and heavily advertised rivals, their market share inevitably declined.

Arthur and May held strong in their belief in producing healthy and nutritious food. They were not prepared to compete with the big corporates if it meant compromising their ideals. Family life was also extremely important to them both, and the expansion needed to compete would have been a compromise too far. In the words of E.F. Schumacher, they believed that 'Small is Beautiful'.

Aside from the threat of competition, Arthur and May had other interests which were taking up much of their time. Arthur was planning a nature trail at Fordhall which would capitalise on the natural advantages of the farm. This was to some extent dependent on Arthur's long-term drainage project which had kept him busy for years. Furthermore, he had been working for some time to develop a revolutionary seed distribution machine which he christened the 'pulvo-seeder'. Unfortunately, their plans were brought to a shuddering halt when May was fatally injured in a road accident not far from Fordhall Way in the summer of 1975.

May's death caused Arthur to reassess his priorities. He closed Fordhall Way

Fordhall Farm *The Yoghourt Years*

1970s Promotional flyers

OUTSIDE CATERING AND SPEAKERS 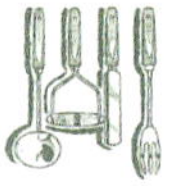
Here the tastes are free and only a small charge is made to cover expenses. We will prepare and provide all that is needed for your Cheese and Wine Buffets and Cocktail Parties. We will see that your party goes with a swing — charge per head to be agreed. Professional speakers can be arranged.

NATURE TRAIL OR FARM WALK 
You are invited, at any time, to go on our planned walk, where notice boards are provided describing each field and the full farming practice explained.
We have two miles of river, a fishing lake, one mile of woodland and a stone-age fortress site. There is a growing population of birds and wild flowers, more are to be brought in for your pleasure. There will be a picnic area for you to relax and enjoy our farm.

FARMSHOP
Every food item prepared or grown at Fordhall is available in large or small packs in our Farm Shop, open seven days a week. On arrival a cup of tea or snack can be obtained, you only have to ask. You may use the enclosed price list to order fresh or for your freezer.

Where we are:
Fordhall Farm, Market Drayton, Shropshire, Tern Hill 255/6

All enquiries to:
Fordhall Organic Farm,
Market Drayton,
Salop. Tel. Tern Hill 255/6

FORDHALL
MENU BROCHURE

Nature is our provider
You are the decider
Our Chef will prepare
Your choice of fare

On our Farm we assist nature to do her job for us by staying in harmony and obeying her instructions; the finest flavours she can offer are revealed through our chef at your table.

FORDHALL ORGANIC FARM
A Fanfare of Joy from Food

GROWN, PREPARED and SERVED

by a small group of Farmers and Producers who care about you and your health. We offer you a range of fresh quality foods that are unique and give a service delivered to your home and freezer. Enjoy planning your party, let us do the worrying. Enjoy shopping at home and let us do the caring.

We shall appreciate getting to know you and you will enjoy the goods and services that are available from our shop.

Visualises chain of farm shops

FRESH FOOD superstores selling farm produce from different parts of the country — this is what a Shropshire farmer visualises and hopes will come about if interested farmers are willing to co-operate.

Mr Arthur Hollins, of Fordhall Farm, Market Drayton, has already opened a farm shop inside his attractive old farmhouse. In the first two months it took £500 which was a good start and he hoped to double this amount during the next two months. As well as selling his own farm products he has on his counter specialities from other farmers — cheeses, free range eggs, fresh vegetables free from chemical sprays, organically-grown stone-ground cereals and wholemeal bread.

He is well known over a wide area as a successful organic farmer and his cream, butter, live yoghurts and lactic soft cheeses have been selling for years in many different parts of Britain.

His 150-acre organic farm, on which he had 40 different species of grass, has Jersey and Ayrshire cattle (which graze out at grass all the year round and which he is now crossing with Herefords and Charolais), sheep, pigs and poultry — and his aim is now to channel everything from the farm back into his shop.

At present he is selling his meat in joints and chops — "because I want people to take them away and taste them for their flavour" — but eventually he plans to sell the meat in bulk to stock deep freezes.

Exchange goods

As well as his dairy products, he makes up farm-house salads, coleslaws, fruit and vegetable salad, fruit salads, cheese cakes and his speciality yoghurts—a yogice (with flavours of lime, chocolate and hazel nuts, strawberries, blackberries, or vanilla and honey) and a yogtail (lemon and advocaat, raspberry and red wine, blackcurrant and rum, lime and lemon, passion fruit and wine, coffee and rum, or mixed fruit). He also sells milk-fed chickens and ducks, bread and dough, brawns and cooked meats.

There are, of course, other farm shops — although not many in the country and Mr Hollins' hope for the future is that an exchange of goods might be arranged between them whereby one van would take produce from his farm shop round to the others and in return bring back their own particular specialities to sell at Fordhall. By doing this only one vehicle would be necessary — to take and bring back — and transport costs would be kept to a minimum.

Cheese-making

"I don't want to form a co-operative, but I think we could help each other in this way and also give the public a wider range of farm produce. I would like to take a load of our goods and bring a load back," Mr Hollins told me. "I am sure the time has come when people prefer fresh farm food and are glad to be able to buy it."

As well as the farm shop Mr Hollins and his wife, Irene May, run several other projects from their farm. They are well used to visitors — 8,000 were there last year — for parties to see round the farm are organised, lectures and talks are given and Mr Hollins personally organises fondue parties for a maximum of 40 people, serving a five course meal which he cooks himself.

Mrs Hollins also runs a guest house, about two miles away from their 14th century farmhouse. It is called Fordhall Way and specialises in wholefood and vegetarian dishes, Mrs Hollins having been a vegetarian for 20 years. Home-made soups, savouries, chutneys and bread, salads, dandelion coffee and herb tea are included in the menus and the only additives Mrs Hollins uses in her cooking are sea salt, cider vinegar, honey, raw sugar, sunflower seed oil and molasses of herbs.

Mr and Mrs Hollins, who have been married for 33 years, run a guest house at their farm (where Mr Hollins' family has been for three generations) 21 years ago when, in those days, they needed to "make ends meet."

Until the war Fordhall Farm was just a cheesemaking concern and Mrs Hollins' late mother was a well known cheesemaker. His son, Robert, who is married with two children, has inherited her ability and makes cheeses for Malbank Farm in Cheshire which in turn supplies the farm shop.

Mr Hollins' father died when he was 14 and he and his mother were left to run the farm — a farm which needed a great deal of hard work because it had become run down. Over the years he studied and practised organic farming and gradually got the farm back on his feet. A great deal of personal research and experiment on his land has given him an enormous amount of specialised knowledge. He often gives lectures and talks to a wide variety of people and he is personally supporting research on his land by students from Keele University who will be looking into plant nutrients and the effect of fertilisers, and the results when none is put on.

As he points out in an article he once wrote for the Soil Association: "We as farmers can only guess at what really goes on and bring results, balancing the needs of all our plants and animals with the elements, assisting nature in her struggle to produce a surplus, and using that surplus for raising our standard of living."

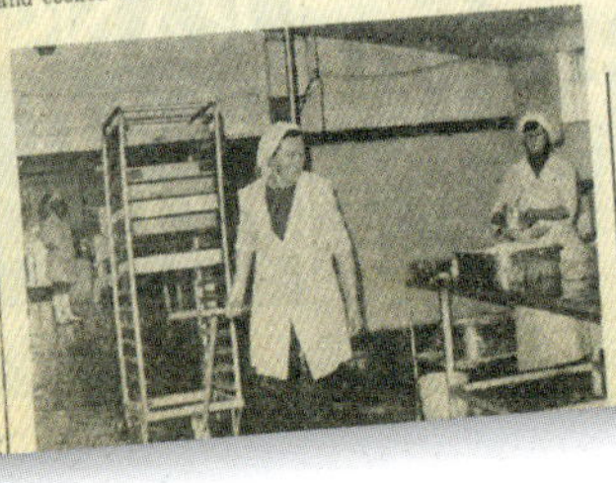

New YFC organiser

SETTLING into her job as YFC county organiser for Brecknock and Montgomeryshire, is Welsh-speaking Ann Davies, of Brynmaing, Uchaf Farm, Pumpsaint, who is enjoying visiting the federation's 23 clubs and meeting all the members.

For eight years she was a member of Carmarthenshire's Dyffryn Cothi club where she was secretary for two years and leader for another two. She has also been chairman of the county girls' committee and the international committee and at Wales level she was vice-chairman of the girls' committee.

She has represented Carmarthenshire in the craft competition at the Royal Welsh Show and was twice in the county team in the Welsh section of the Wales public speaking contest.

In 1972 she went to Canada, as a guest of Ontario Junior Farmers and enjoyed seeing the country as a member of a farming family rather than as an ordinary tourist.

The 120-acre family farm at Pumpsaint

German theme

THERE was a German theme to the annual produce rally of the Shropshire Federation of Women's Institutes in Shrewsbury last week, when more than 300 members saw a film on German wine and heard a talk by Dr Edgar Gerwin on the country's food and wine. Afterwards German sausages with bread, sauerkraut and gherkins were served with German wines.

May 1975, Farmers Guardian. Arthur and May were looking to expand their local food business through the creation of a farmers' cooperative. May's tragic death brought this plan to a shuddering halt soon after. Interestingly, the concept bears many similarities to today's local food movement

64   Fordhall Farm *The Yoghourt Years*

May 1975 Wolverhampton Chronicle

and returned the catering operation to the farm. Shortly afterwards, a serious fire in one of the store rooms sealed the fate of the dairy revolution at Fordhall. Arthur gradually wound down the dairy business with the final sale of his equipment and machinery in March 1979.

Arthur went on to develop his organic beef suckler herd, working with independent butchers to get his products direct to the customer. He continued to promote his farm trail and farmhouse meals, as well as advertising working holidays, maintaining links with the land and the people just as he always had. And in any spare moment he had, he wrote his autobiography in the 1980s, **The Farmer, the Plough and the Devil**, charting the struggle to turn around the fortunes of this little bit of Shropshire countryside.

Yoghourt was never to be made at Fordhall commercially again. The irony was not lost when a well-known German 'yogurt' manufacturer moved next door in 1991, creating a new fight for survival for Arthur Hollins and his family.

Arthur spent many hours working with Terry Jakeman draining his water meadows

Mother Nature was their guide

1993 Arthur with some of his treasured little workers

Arthur and May proved to be an unbeatable team; this was largely due to their inbuilt tenacity and drive, but what set them apart from the rest was their combined love and appreciation for Mother Nature.

Arthur's love was for the earth, wildlife and farming, whilst May's passion for the relationship between food and human health completed the cycle. Together, they understood the critical importance of diversity and balance; this awareness shone through every aspect of their business.

Arthur's understanding came after taking over the ruined farm from his father. Arthur and his mother had had a huge task ahead of them, but learnt much from the old farm workers. They taught Arthur about the importance of real manure for the soil in comparison to the chemical fertilisers his father, Alfred, had used. His interest was stimulated, the benefits were obvious and here his life's journey began.

Arthur's eldest daughter recalls the passion that both of her parents had for nature:

> "It's important to keep a balance of beneficial bacteria in your stomach and yoghourt encourages this. They [Arthur and May] were passionate about helping people have healthier lives. Their whole life's philosophy was using nature's surpluses and making the most from their natural environment. Dad was very well read... he would read avidly in the evenings; despite leaving school at such a young age, he was very enquiring about the science of farming.... He believed if you worked with nature instead of against it, nature will serve you in bounty.

Fordhall cattle take their pick from over 70 plant species in the farm's pastures

Fordhall Farm *The Yoghourt Years*

1993 Starlings at Fordhall

Arthur's system of rotation and grass species allowed his livestock to remain outdoors year round

"Mum's philosophy was exactly the same... she believed food should be consumed in as natural a state as possible, and of course making it very tasty using natural ingredients ... They believed in trying not to use artificial additives in food or in the soil, and they believed if you altered natural foods too much, this would impact on your body."

Arthur's philosophies for health, experimentation, and improvement continued at Fordhall throughout the years of the dairy. His outdoor grazing system, the rotations used and the types of grass and forage planted are extensively reported in many magazines of the time including **Farm and Country Magazine**, 1961 and **The Listener** in 1971. His constant experimentation, matched with exhaustive reading and research, allowed Arthur to create his own unique way of farming.

This eventually led him to a system of outdoor grazing as an article in the **Delicatessen** from 1965 explains *"Unlike most dairy farms, Fordhall keep their cows out all year round; fresh grazing is continued by the planting of special pastures which are rested during the usual growing season and bear grasses which can continue to grow at exceptionally low temperatures."*

Arthur found this system to be healthy for his pastures and for the livestock. He saw no reduction in milk production, and his cows were happy. His research into protecting the soil continued, and he developed a cultivating machine patented as the Pulvo-Seeder, and later known as the CulturSeeder. This machine maintained a soil mulch cover whilst sowing seeds below it. It aerated the soil and did not cause compaction or destroy the soil layers. Arthur believed this was the future for sustainable arable farming.

> "We as farmers can only guess at what really goes on and judge from what appears to bring results, balancing the needs of all our plants and animals with the elements, assisting nature in her struggle to produce a surplus, and using that surplus for raising our standard of living." **Arthur Hollins, Healthy Living, April 1975**

> "When I came for my interview, he took me into the field to the left and there was a lovely herbal lay – I'd never heard of it before – there was chicory and yarrow and all these herbs. They were so deep rooted and they'd be bringing minerals coming up and that in turn gets into the cows and he was so enthusiastic over it. It looked delicious!"
> **Chris Clowes**

Unfortunately, after the dairy closed in the late 1970s, Arthur found it difficult to find the time to get the machine to market. Nevertheless, his unique system of outdoor grazing, based on rotation of livestock and a diversity of plants in the pastures, remains at Fordhall to this day, and is maintained by his youngest son, Ben Hollins.

> We used a crop of hardy greens, rape and rye, with the mushrooms and farmyard manure spread on top after the crop was up...
>
> This method provided cover for all the cold-blooded animals of the soil: they in turn, cultivated that soil in the root areas of the plants. It provided darkness and humidity for the vigorous growth of fungi, producing large quantities of nitrogen, and kept the eelworm and wireworm down to balanced numbers by trapping them. It gave nature a chance to create an ecological balance of insects and soil-animal life, their only way of evolution, through a survival of the fittest, with all of them depending on each other for food. The plants returned vast quantities of living cells created through photosynthesis from above the soil, to add to the soil micro-billions. Amino acids were fed back into the root area by the plants all winter; every sunny day gave a bonus. Feeding animals on such fields meant that the remainder of the crop was being composted by the animals' stomachs and the surplus energy returned to the soil by their droppings and urine, which encouraged visits from thousands of birds: they too left a bonus...
>
> I was beginning to learn how to live off nature's surplus, and to understand that nature's culti-vating machinery – the soil animals – was far superior to anything man could devise. Organic farming, to me, means using an area of land to grow plants, to be used as food for other human beings, and availing myself of all the tricks of the trade which nature has developed during the billions of years it has taken her to create soil.
>
> **Arthur Hollins,**
> The Listener, *2 December 1971*

Footnote: More information on Arthur's organic farming practices, research and philosophies can be found in his book *'The Farmer, the Plough, and the Devil'*. Available from Fordhall Farm.

The new fight for survival

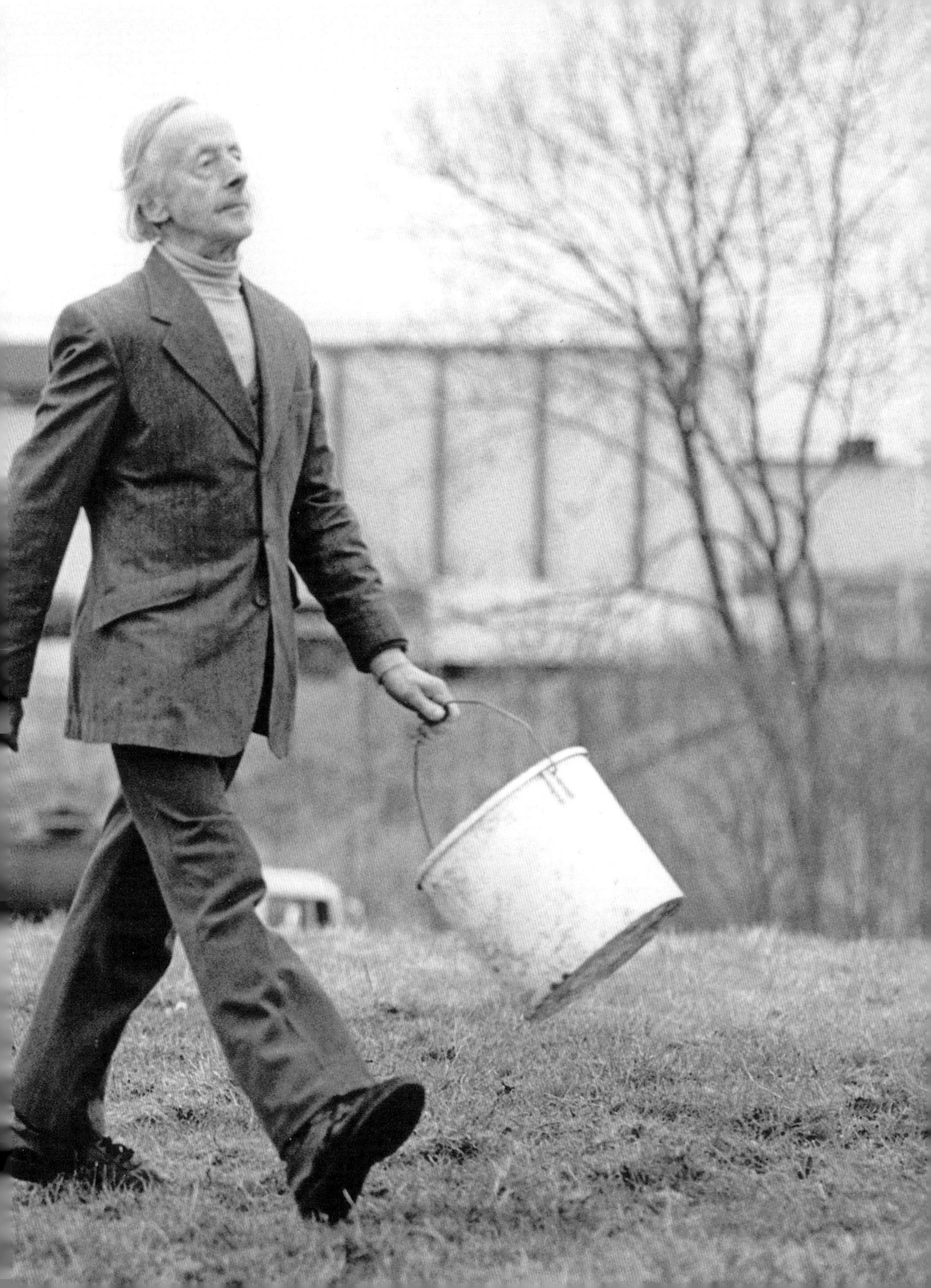

The new fight for survival

After May's death in 1975, Arthur worked hard to ensure the farm remained busy and full of people. As well as farm trails, a farm shop, restaurant meals, school visits and tours, he advertised working holidays.

Connie Trojanski had seen a yoga retreat at 'Fordhall Way' advertised in a free health newspaper in the Bradford Holland and Barrett store, and after sampling the Fordhall yoghourts in another local health food shop, decided that this holiday suited her healthy living ethos. Unfortunately, in the wake of May's tragic death, Connie was advised that the holiday would have to take place at Fordhall Farm with good food and day trips, but without the yoga. Connie thoroughly enjoyed her time at the farm and wholeheartedly believed in Arthur's ethos. She returned a few years later for a working holiday and never left. Connie married Arthur in 1981, and soon after followed Arthur's second generation of children, Charlotte and Ben. The organic way of living clearly kept him sprightly, being 67 years of age when Charlotte was born and 69 when Ben was born.

Within a decade, a large German 'yogurt' manufacturer had moved onto land adjacent to Fordhall, bringing a new challenge for the Hollins'. Although the family had farmed the land for generations,

> "If there is one thing dad absolutely loved, it was having lots of people coming to the farm. That was how he felt the message came over best, by people actually being at the farm."
> *Marianne Howe, née Hollins*

Connie and Arthur married in 1981

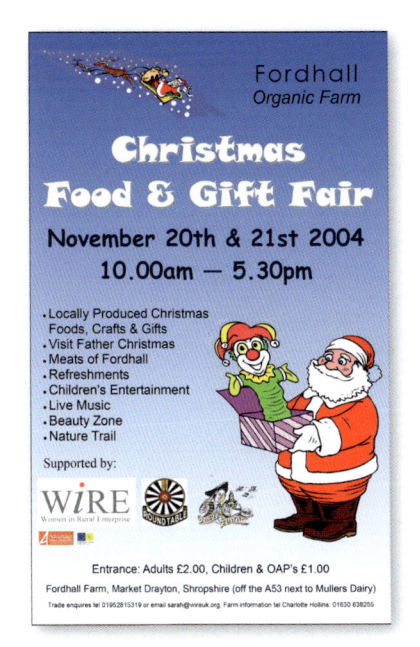

2004 Christmas Fair

they had remained tenant farmers, and with the new neighbours, the landlord saw an opportunity to sell Fordhall Farm for industrial development. Arthur was once again faced with the power of the corporate yoghourt world.

A 15-year legal battle with the landlord ensued. Each year more and more vital funds were used to pay legal costs; reinvestment within the farming enterprise ground to a halt. Eventually, livestock and machinery had to be sold to cover the incoming bills.

With the farm deteriorating, Arthur's advancing age, and Charlotte and Ben still at school, the odds were stacked against them, but Arthur would not let go of the legacy he had spent his life creating. As the business shrank through the 1990s, staff numbers fell to just one: stockman Terry Jakeman. Even he was made redundant in 1996 when the livestock levels reached an all-time low and there was no further use for his expertise.

After many generations of tending the land at Fordhall, the final eviction notice was issued and the family were due to leave the farm forever in March 2004. Recently graduated from university and college, Charlotte and Ben at 21 and 19 years respectively, returned to Fordhall and took up the fight themselves.

Remarkably, thanks to an ease in development pressures, the siblings obtained an 18-month tenancy agreement only 24 hours before they were due to be evicted. With the support of an ageing Arthur they began to sell their produce direct from the farm and quickly realised themselves the importance of people to Fordhall. In November 2004 they held their first event, a Christmas Food and Gift Fair. It was a huge success and although Arthur was weak due to his advancing age, he loved to see people back at Fordhall again. Sadly, he passed away only two months later, in January 2005, in his 90th year.

"A friend and I came to the farm and had a wonderful holiday, going out on day trips and eating so much delicious home cooking – I put on half a stone! I came back on working holidays to see how it was run behind the scenes, and Arthur was still bringing yoghourt up to stores in Bradford. Eventually, I returned to Fordhall at Christmas 1979 and I am still here 35 years later. It was so busy with the farmhouse restaurant in the late 1970s, there would regularly be 50-60 guests for an evening meal on a Saturday night – the whole house was full. I remember lots of laughing and joking and even dancing on the tables!"
Connie Hollins née Trojanski.

Fordhall Farm *The Yoghourt Years* 73

The Observer 9 October 2005

Yours for £50 – a slice of Britain's rural heritage

Fordhall Farm's shop, also selling the produce of ci... animals and its own yog...
Photographs by Andrew

Green farm in £1m fight to survi...

First 'organic' farm is under threat from developers

by Juliette Jowit
Environment Editor

CHARLOTTE HOLLINS surveys the fields patched with nettles, dock leaves and thistles with a look of pride. 'A lot of farmers would look at this and say it's untidy, but it's a live farm and it's sustainable,' she says.

For all its scruffiness, this little patch of Shropshire is a piece of English landscape with a remarkable history, one of the first to practise the principles of organic farming and produce yoghurt. Now, in a tale all too familiar the length and breadth of Britain, the farm is under threat from developers. The difference here is that a remarkable alliance of neighbours, local businesses and wildlife-lovers has risen up in support of a £1 million bid to save the farm for the nation – at £50 a share, the public...

...ated one of the original 'organic' farms – although it was never officially certified.

'Dad never liked the word organic; he thought this is how farming should be, it's just nature,' says Charlotte...

...education and encouraging people to feel welcome to visit, work or just explore the farm and its teeming wildlife.

They have until next summer to raise more than £1m – £800,000 to buy the farm, the remainder to set up the education and community pro...

bypasses and airpor... the invisible damag... tion, chemically int... farming and traffic...

All these are eati... area of countryside... of Southampton eve... claims a report by t... paign for the Protec... Rural England. 'Th...

Councillors object to tower plan

Parish councillors in Telford have come out against moves to put up a 50ft telecommunications tower in the town.

Hutchison 3G has submitted plans for the structure at the TA Centre, Territorial House in Trench Road, Trench.

It would include three antennas, two dish antennas, radio equipment housing, and a 2.1m high palisade fencing.

Members of Lilleshall and Donnington Parish Council's finance committee resolved to object to the application at a meeting.

They also resolved to make no objection to a two storey home with a garage on garden land at the rear of 1 Barrack Lane, Lilleshall.

There was also no objection to plans to demolish a chimney, erect a new flue and convert an existing store and woodship store at Lilleshall Primary School, Limekiln Lane.

Police alerted to town fighting

Police were called to an area of Newport after reports of fighting involving about 30 youths.

They went to Broadway in the town at about 7.25pm on Friday but although there was a large group of youths there was no fighting.

Police planned to visit a number of people over the weekend to speak to them about their conduct.

Pensioners off on trip to coast

More than 200 senior citizens in Telford are set to enjoy an outing to the coast

---

Jump to it – the siblings are now hoping to transform Fordhall

Just desserts as chef wins award

Telford chef Terry Bowden not only won a gold medal for his custard tart at a cooking competition he beat 30 oth...

...naire which saw Sodexho employees compete in 12 different categories from dessert tart, plated main course, to decorated novelty cakes and innovative...

Terry said: "I was really pleased to get gold for my entry and for it to be chosen as the best in class was fantastic. In the programme it beat desserts...

The chairman of Market Drayton food group has spoken of his delight after an unique organic farm was saved from developers and preserved for the community.

After years of fighting to save their family home, Ben and Charlotte Hollins, confirmed Fordhall Farm, at Tern Hill, had been saved just days before a deadline was due to lapse on Saturday.

The brother and sister had to raise £800,000 but with the help of widespread funding and celebrity backing the organic site was secured and it is now community-owned as a trust.

The family who will stay on at the farm will now transform Fordhall into an educational and social resource centre helping to reconnect people and children to food and farming.

Determined

Taste of the Town chairman Jeff Hopkins, whose daughter Sophie is project manager at Fordhall, said the energy and commitment shown by everyone was a reminder that passion and purpose were alive in rural Shropshire.

The food group which helps to bring producers together and promotes produce from the area is hosting a giant food exhibition at the Grove School from July 5 to 8.

Mr Hopkins said: "The Hollins family are determined to protect and promote Fordhall Farm as an example of the way that land can be worked sympathetically with nature."

Ben and Charlotte's father, the late Arthur Hollins, stopped putting artificial fertilisers and pesticides on the land at Fordhall over 65 years ago and spent his life nurturing the soil.

The family sold £50 shares in the farm to help raise the funds to save it and other cash came from grants and fundraising events

The picturesque farmhouse at Fordhall

Messages sent after farm saved

Messages of support and congratulations have been sent from across the UK and abroad to everyone involved in saving Fordhall Farm.

Ben and Charlotte Hollins, of Fordhall Farm, at Tern Hill, have been flooded with messages of support from across the country and Europe following their success.

The farm has now been bought.

It is in the hands of the Fordhall Community Land Initiative and is owned by residents who bought £50 shares.

The campaign attracted strong international support and one message says the family and workers "organised a miracle" to raise more than £800,000 and save the farm from developers.

Fordhall's fight made the national press

74   Fordhall Farm *The Yoghourt Years*

Undaunted, Charlotte and Ben continued the fight to save their family heritage and ensure this land remained an asset to the community for generations to come. They secured an option to buy 128 acres of Fordhall Farm in July 2005. A community share issue was launched and the siblings captured the imagination of the British public with the plight of their family home. £800,000 was raised in only six months through an innovative £50 community share issue. After months of hard work and late nights, ably supported by huge numbers of the local community, Fordhall was heralded England's first community-owned farm in July 2006.

Ben Hollins, Arthur's youngest son, now farms Fordhall on behalf of 8,000 international community shareholders. They are his landlord via the Fordhall Community Land Initiative (the industrial and provident society that now owns Fordhall Farm). Ben manages the land organically, maintaining Arthur's unique Foggage farming system, whereby livestock remain outdoors all year round, eating only grass, just as nature intended.

Bringing Fordhall into community ownership in 2006 was a pioneering move, but Arthur and May would both have been familiar with much of the activity that goes on at the farm today. There is a farm shop, a restaurant (café) serving local and organic delights, working (volunteer) holidays, educational visits, glamping facilities, guided tours, farm trails and events. Only the Jersey herd and the bustling dairy no longer feature in the Fordhall picture.

The concept of homemade local produce without artificial additives and preservatives still underpins Fordhall's philosophy, as does its close connection with its customers and the local community. Today, just as in Arthur and May's glorious yoghourt years, everyone is welcome at Fordhall.

In some ways things don't change, they just look a bit different!

2004 Arthur with his youngest son and daughter, Ben and Charlotte

Fordhall Farm *The Yoghourt Years*   75

Above: Ben's Aberdeen Angus and Hereford beef herd remain 100% grass fed on Fordhall's organic pastures. All beef, pork and lamb produced on the farm continues to be sold direct to the public
Below: The Old Dairy building was transformed in 2011 using eco materials; it now houses the farm shop, café and meeting room space

Footnote: More information on Fordhall's fight for survival against industrial development can be found in the book by Ben and Charlotte Hollins, ***The Fight for Fordhall Farm***.

Fordhall Farm Timeline

1915 Arthur Hollins born 20th May 1915

1928 Arthur's father Alfred Hollins dies on 29th June 1928 at the age of 38 years

Arthur takes over the tenancy at Fordhall Farm at the age of 13 years and 1 month

1942 Arthur marries May Baker

1949 Arthur decides to sell shorthorns and buy Jersey cows

Farmhouse is run as Tern Hill Country Club and guest house to raise funds for the Jersey herd

1951 Fordhall herd of pedigree TT Jerseys founded

1954 End of national rationing and food production restrictions

1956 Milking machine adapted to milk four cows at once, carrying milk direct to the separating chamber

Clotted cream and cream cheese first made and sold

1957 Yoghourt and yoghourt cheese first made and sold

Jersey herd is first kept outside all year round

The Shippon is turned into a modern milking parlour

Large deep-freeze unit installed

Customer visiting club established

1958 Cottage cheese first made and sold

1961 First full-time delivery driver is employed

1962 Fordhall exhibits at Olympia Food Fair, London

1963 Swiss-invented Ski Yoghurt is launched in the UK

1963 Herringbone milking parlour built and installed, shippon is turned into the new dairy factory

1965 Fordhall advertises in Australia and New Zealand for dairymaids

1966 Fordhall employees visit Granada Studios and Manchester United Football Club

1967 Yogice and Yogtails first made and sold

1970 Arthur patents his Pulvo-seeder

1971 60 varieties of dairy products are now being marketed

Keele University uses the farm for research linked to Arthur's farming philosophies

1972 First farm shop opens

Fordhall Way opens at Pell Wall, Market Drayton

1973 Arthur invents and patents his display refrigerator

1974 First vegetarian wedding breakfast at Fordhall Way

1975 May Hollins dies

1979 Dairy finally closes and all remaining machinery is sold

1981 Arthur marries Connie Trojanski

1982 Charlotte Hollins born

1984 Ben Hollins born

2004 Ben and Charlotte Hollins take over the Fordhall tenancy agreement

2005 Arthur Hollins dies, four months before his 90th birthday

2006 Fordhall Farm is placed into community-ownership. Ben Hollins continues as tenant farmer. Charlotte Hollins manages the Fordhall Community Land Initiative (the new landowner)

About our oral history contributors

Jill Blud née Comben
Worked in the dairy at Fordhall and lived in the farmhouse from January to September 1966 before going on to agricultural college.

Helen Burgess
Started working in the chillers around 1973 on Sundays and during school holidays. Later progressed on to making yogice. Worked at Fordhall for approximately three years.

Chris Clowes
Trained at Rodbaston Agricultural college and lived and worked at Fordhall tending and milking the cows for about a year, 1963-64.

Mary Cowen
May's niece. Holidayed at Fordhall and on leaving school at 18 worked as herdswoman until she left to look after her father. Returned to the farm to work as a driver and demonstrator until 1968.

Eric Davis
Became an independent distribution agent for Fordhall in the early 1970s. Established a cold store in Barlaston.

Pat Davis
Married to Eric. Worked with him as a Fordhall distributor.

Geoff Fletcher
Started in August 1969 as assistant herdsman. Later progressed to Farm Manager until March 1971.

Pete Guildford

Worked as a store man from around 1968 to 1971, also worked as a relief milker and occasional yoghourt maker.

Terry Healey

Started work at the farm shortly after leaving school in the early 1960s. Had a variety of jobs including working as sales demonstrator travelling to stores, markets and exhibitions all over the country.

Connie Hollins née Trojanski

First came to the farm in 1975 to take advantage of the yoga holidays at Fordhall Way. Married Arthur in 1981 and has lived and worked at Fordhall ever since.

Terry Jakeman

Lifelong resident of north Shropshire. Began work in the dairy in the 1970s and although he has been employed elsewhere, Terry has spent most of his working life managing the livestock at Fordhall, where he still works alongside Ben Hollins as stockman.

Jean McAdie

Came to the farm in 1963 to work on the printing machine. Remained at Fordhall for about a year.

Mike Niccolls

Employed at the farm aged 18 in 1961. Stayed for 18 months, milking, delivering and demonstrating Fordhall products.

Pam Price

Worked as a delivery driver for Fordhall products in the London area in the 1960s and 1970s.

Fordhall Farm *The Yoghourt Years*

Acknowledgements

This book is only possible because of the generous grant received from the Heritage Lottery Fund. This covered the cost of archiving, cataloguing, researching, editing, designing, managing and publishing the material within this commemorative book, as well as that held at Fordhall Farm and Shropshire Archives.

Fordhall Farm The Yoghourt Years was brought to you after many months of research and hours of work from a small group of committed volunteers. They worked closely with Charlotte Hollins to gather the facts, check the references, write and edit text, and tirelessly trawl through hundreds of photographs. Without their dedication and commitment to the project, the information contained within would remain scattered throughout the Fordhall farmhouse and inside the hundreds of pages of text held at Shropshire Archives. It is with huge gratitude that we thank Gary Kanes, Viv Watkins and Rob Woodward.

Much of the text and the quotes came from the many people who offered their memories through oral histories. These were recorded with the help of volunteers Alan Howe and Sharon Jones, as well as experienced oral historians Chris Eldon Lee and Genevieve Tudor.

For the design, editing and publishing thanks go to Mike Ashton, Heidi Robbins and Ina Taylor, and to Ruth King for assisting with proof reading.

Many of the photographs used within the book are republished thanks to a number of periodicals. Those we have been able to identify are credited in the photo credits at the end of the book.

We would like to thank all those who are directly quoted in the book. These are Jill Blud, Helen Burgess, Chris Clowes, Mary Cowen, Eric Davies, Pat Davies, Geoff Fletcher, Pete Guildford, Terry Healey, Marianne Howe, Terry Jakeman, Jean McAdie, Mike Niccolls, and Pam Price.

We would also like to thank all those who may not have been quoted, but whose memories helped us to get an all round picture of Fordhall Farm fifty years ago. Thanks to Bill Ashley, Harry Bates, Jean Burgess, Paul Burgess, Claire Cope, Carol Edmonds, Bill Edwards, Frank Fuller, David Geoffries, Margaret Gibbons, David Gordon, Colin Hill, John Holmes, A Howell, Liz James, John KirkPatrick, Pat and Paul Raymer, Isobel Thomas, Helen Taylor (née Worth), Arminal Trueman, Jean Watkins and Colin Woolley.

None of this would have been possible without the help and cooperation of Arthur's second wife, Connie Hollins. She has pulled out every news cutting and photograph she could find, helped us check facts, and offered useful insights, as have other members of the Hollins family. The team at Shropshire Archives have gathered documents, carefully conserved those which were damaged, catalogued them and now made them available for public viewing. The work of the Archives and their committed team of volunteers has made this huge resource accessible to us all for the foreseeable future.

We apologise if we have missed anyone from this list – it has been a big Project over many years and it is still ongoing. If you too have a memory to share, if you used to work at Fordhall or spent a holiday there, please do get in touch – we would love to hear from you.

Lastly, thank you for picking up this book and reading it. We hope you enjoy it and you learn something new. Maybe we'll see you at Fordhall, helping to create the next piece of history!

Credits and References

The editor and publisher gratefully acknowledge the permission granted to reproduce the copyright material in this book, courtesy of the Observer, Shropshire Star, and Staffordshire Sentinel News & Media.

Australia & New Zealand Weekly. "*Attractive Employment for Australians and New Zealanders*". July 10 1965. Advertisement page 49.

The Chronicle. *Title unknown,* September 9 1971. Image page 60 top.

Delicatessen Magazine. "*Now Fordhall offer a salad to go with their 'Cream'* ". July 1975. Pages 12, 68.

Evening Sentinel. "*People beat a path to his door*". January 15 1975. Image page 25.

Farm & Country Magazine. "*Cream, cheeses and yoghourt from a Shropshire dairy herd.*" June 6 1961. Images page 13 top and bottom, 16, 32 bottom, 43. Page 68.

Farmers Guardian. "*Visualises chain of farm shops*". May 9 1975. Page 57, 64.

Farmers Weekly. *Date and title unknown.* Images page 19 bottom, 50 top right.

Farmers Weekly. *Title unknown.* April 14 1972. Image page 63.

The Guardian. *Title unknown.* 1993. Images pages 66, 70-71 and back cover.

Healthy Living. "*Healthy living down on Fordhall Farm*". April 1975. Page 68.

Hollins, A. "*The Farmer, the Plough and the Devil*". Ellingham Press. 2006. ISBN 0-9547560-3-7. Pages 8, 65.

The Listener. "*Fordhall's – Arthur Hollins tells the story of his organic farm*". December 2 1971, Vol. 86, No. 2227. Pages 22, 37, 68, 69.

Liverpool Daily Post. "*Healthy enterprise that suits the public taste*". June 8 1971. Page 49.

Newcastle Advertiser. *Title unknown.* November 1993. Image page 62.

Newport Advertiser. "*Visit to T. V. Studio*". March 15 1974. Image page 64. Page 55.

Newport Advertiser. "*Cleanliness is the Keyword at this Market Drayton Farm*". March 5 1965. Image page 44 top right.

Newport Advertiser. "*Australian girls will soon be here: From "Down Under" to Market Drayton Farm*". Jan 22 1965. Image page 46 top left.

Newport Advertiser. *Fordhall Way Advertisement.* March 15 1974. Image page 61 middle.

The Observer. "*Green farm in £1m fight to survive*". October 9 2005. Image page 74.

Shropshire Star. "*Getting away from the pre-packed foods...*" February 4 1966. Front cover image.

Shropshire Star. *Date and title unknown.* Image page 13 top left.

Shropshire Star. "*Success of farm bid delights food chief*". July 3 2006. Image page 71 bottom.

Shropshire Star. *Date and title unknown.* Image page 69 bottom left.

Soil Association magazine, Living Earth. "*Organic origins: Next of kin*". Summer, 2005. Image page 75.

Staffordshire Sentinel. *Date and title unknown.* Image page 2.

Staffordshire Sentinel. "*Big business on Salop farm*". Date unknown. Page 9.

Staffordshire Sentinel. *Date and title unknown.* Image page 17 bottom.

Staffordshire Sentinel. *Advertisement.* Thursday 7 1957. Image page 37 bottom.

Staffordshire Weekly Sentinel. "*Seventy-one varieties of dairy products come from Fordhall Farm*". September 24 1971. Image page 22.

Sunday Pictorial. "*Her cheese comes from the farm in 40 flavour*". 1958. Image page 10.

Wellington Journal & Shrewsbury News. "*Farmhouse Yoghourt Being Made in Shropshire*". November 16 1957. Images page 12, 15.

Wellington Journal & Shrewsbury News. "*The Bird Cage*". August 14 1964. Image page 52.

Wellington Journal & Shrewsbury News. "*Shotgun Wedding*". Date unknown. Image page 53.

Wolverhampton Chronicle. "*A message from the 'muck and magic' farmer*". May 22 1975. Image page 65.

Photographers

Oliver Allan. Images page 76 top and bottom.

Liz Kessler. Images pages 51, 65 bottom, **68 top left**.

Melanie Rothner. Image page 82 right

Lisa Perry. Images of all pots and lids throughout the book.

Shropshire Archives

www.shropshire.gov.uk/archives

8035/2/2/4/2-4 Book of credit notes. Page 29.

8035/2/2/4/3 Accounts for goods supplied to Caldey Island Estate Company. Page 34.

8035/6/4/2/1-13 Correspondence regarding planning permission for proposed extension to factory. Page 46.

Every effort has been made to trace copyright holders and to obtain their permission for the use of copyright material. The publisher apologises for any errors or omissions in the above list and would be grateful if notified of any corrections that should be incorporated in future reprints or editions of this book.

Top: 2006 People continue to be at the heart of Fordhall
Middle: 2013 Ben and wife, Marie-Anne, with son Jamie Arthur Hollins
Bottom: Charlotte and Poppy contemplate Fordhall's future